State-of-the-Art of Mathematical Modeling, Dynamics, and Control of Wind Turbines Engineering

Edited by José M. Balthazar, Ângelo M. Tusset, Clivaldo de Oliveira and Raphaela Carvalho Machado

Published in London, United Kingdom

State-of-the-Art of Mathematical Modeling, Dynamics, and Control of Wind Turbines Engineering
http://dx.doi.org/10.5772/intechopen.1003403
Edited by José M. Balthazar, Ângelo M. Tusset, Clivaldo de Oliveira and Raphaela Carvalho Machado

Contributors
Angelo M. Tusset, Athanasios Kolios, Azeez Olasoji, Bruno Stuyts, Carlos Sastra Jurado, Christof Devriendt, Clivaldo de Oliveira, David Oyedokun, Jeferson Jose de Lima, José M. Balthazar, Komla Folly, Lin Wang, Mkhutazi Mditshwa, Natasha Martins Rodrigues de Jesus, Raphaela Carvalho Machado, Selahattin Ozcelik, Srikanth Bashetty, Teófilo Miguel de Souza, Thais Santos Castro, Wout Weijtjens

First published in London, United Kingdom, 2025 by IntechOpen
IntechOpen is the global imprint of INTECHOPEN LIMITED, registered in England and Wales, registration number: 11086078, 167-169 Great Portland Street, London, W1W 5PF, United Kingdom

For EU product safety concerns: IN TECH d.o.o., Prolaz Marije Krucifikse Kozulić 3, 51000 Rijeka, Croatia, info@intechopen.com or visit our website at intechopen.com.

British Library Cataloguing-in-Publication Data
A catalogue record for this book is available from the British Library

State-of-the-Art of Mathematical Modeling, Dynamics, and Control of Wind Turbines Engineering
Edited by José M. Balthazar, Ângelo M. Tusset, Clivaldo de Oliveira and Raphaela Carvalho Machado
p. cm.
Print ISBN 978-1-83769-961-2
Online ISBN 978-1-83769-960-5
eBook (PDF) ISBN 978-1-83769-962-9

If disposing of this product, please recycle the paper responsibly.

Meet the editors

Prof. José Manoel Balthazar is a distinguished scientific researcher specializing in nonlinear dynamics, chaos, control, and their applications to engineering, including science and experimental aspects of nonlinear dynamics and Electromechanical Systems across MACRO, MEMS, and NEMS scales. He holds a Faculty of Mechanical Engineering professorship at São Paulo State University (UNESP) in Bauru, São Paulo, Brazil. Additionally, he serves as a Visiting Professor in the Department of Electrical Engineering at the MEETFederal University of Technology of Paraná (UTFPR) in Ponta Grossa, Paraná, Brazil. Prof. Balthazar is a full member of the Academy of Sciences (ACIESP) in São Paulo, Brazil, and a Fellow of the CNPq-Brazilian Council for Mechanical Engineering Research. He earned his Ph.D. in Mechanical Engineering from the University of São Paulo in São Carlos, SP, Brazil. His expertise extends to serving as a journal reviewer in nonlinear dynamics, chaos, and control, further highlighting his significant contributions to these research areas.

Angelo Marcelo Tusset was born on 14. 04. 1969 in União da Vitória in Brazil. He received a bachelor's degree in Sciences from the State Faculty of Philosophy Science and Letters (FAFI) in 1995 (Brazil), a bachelor's in Mathematics at the State Faculty of Philosophy Science and Letters (FAFI) in 1996 (Brazil), and a bachelor in Engineering of Automation and Control in the UNC in 2007 (Brazil). He received his master's in mathematical Modeling at UNIJUÍ (Universidade Regional do Noroeste do Estado do Rio Grande do Sul) in 2004. Afterwards, he received his Ph.D. in Mechanical Engineering at the UFRGS (Universidade Federal do Rio Grande do Sul) in 2008. He was a post-doctoral fellow in mechanical engineering at UNESP (Universidade Estadual Paulista) in 2010. He has been an Associate Professor of the UTFPR (Federal University of Technology – Paraná) (Brazil) since 2010.

Professor Oliveira holds a doctorate in Mechanical Engineering from the State University of Campinas (2003), a master's degree in electrical engineering from the same institution (1997) and a bachelor's degree in Physics from UNESP (Universidade Estadual Paulista) in 1993. He was director of the Faculty of Engineering at the Federal University of Grande Dourados/MS from 2011 to 2015 and is currently a full professor there. He has extensive experience in Mechanical Engineering, with an emphasis on automation, and works mainly on the following subjects: control, non-linear mathematics, robotics and mechatronics. He did post-doctoral work at ITA in the Mechanical Engineering department, focusing on Modeling and Identification of Dynamic Systems.

Raphaela Carvalho Machado currently serves as an Assistant Professor in the Department of Electrical Engineering (DEE) at UNESP Campus Guaratinguetá. She graduated as an Industrial Electrical Engineer from the Federal Institute of Maranhão (IFMA) in 2011 and obtained her Master of Science degree in Electronic and Computer Engineering in 2013, followed by a Ph.D. in Aeronautical and Mechanical Engineering in 2019, both from the Technological Institute of Aeronautics (ITA). She is particularly interested in theory-based research and teaching activities that apply system identification techniques to experimental data, control systems, flexible aircraft dynamics, structural dynamics and modal analysis, electronic circuit design and measurement and data acquisition systems.

Contents

Preface

Wind turbines are devices that convert the wind's kinetic energy into electrical energy. They are one of the most promising renewable energy sources, as they have minimal environmental impact and can operate in various locations and climates. There are many types of wind turbines, each with different characteristics and advantages. Some of the main categories are horizontal axis turbines, vertical axis turbines, offshore turbines, small-scale turbines and others. Wind turbine models are designed to optimize turbine performance, reliability, and efficiency, considering wind conditions, site characteristics and grid requirements. Wind turbine models can also be used for simulations, testing, and research.

This book contains five chapters organized into two sections.

Acknowledgements are due to all the authors, co-authors, and reviewers for their time, collaboration, interest, and enthusiasm in supporting this book.

The editors also acknowledge Mrs. Marica Novakovic at IntechOpen for her help throughout the publication process.

José M. Balthazar
UNESP - Bauru,
SP Brazil and Federal University of Technology Paraná,
Ponta Grossa, Brazil

Angelo M. Tusset
Federal University of Technology Paraná,
Ponta Grossa, Brazil

Clivaldo de Oliveira
Federal University of Gran Dourados,
Dourados, MS, Brazil

Raphaela Carvalho Machado
UNESP, Campus Guaratinguetá,
Guaratinguetá, SP, Brazil

Section 1

Control and Optimization Approaches

Chapter 1

Integration of Wind Power Plants for Power System Frequency Stability Improvement

Mkhutazi Mditshwa, Azeez Olasoji, Komla Folly and David Oyedokun

Abstract

The integration of wind power plants (WPPs) into modern power systems presents both opportunities and challenges, particularly in ensuring power system stability and frequency stability. This chapter explores the methodologies, technologies, and challenges associated with leveraging WPPs to enhance power system frequency stability. It begins with an overview of the operational characteristics of WPPs and their impact on power system dynamics. Mathematical models and simulation techniques, focusing on the IEEE 9-bus network as a case study, are discussed to illustrate the dynamics and control strategies necessary for effective integration. This study focused on developing a supplementary control loop (secondary control scheme) to enable WPPs' participation in frequency regulation. The efficacy of the developed control scheme was evaluated to assess the contribution of the WPPs to frequency regulation. DIgSILENT software was used to model the IEEE 9-bus network and evaluate the control scheme performance. The chapter also presents detailed case studies involving load demand increases and generation loss scenarios to demonstrate the practical application and effectiveness of the proposed technique. The results showed that WPPs can participate in secondary frequency control, and this will limit the reliance on fossil fuel-based conventional generators for ancillary reservices.

Keywords: wind power plants, frequency stability, control schemes, renewable energy integration, power system stability, wind turbine control

1. Introduction

There is a global push to transition toward more sustainable methods of electricity generation. Therefore, power systems across the globe are gradually displacing synchronous generators (SGs) with renewable energy sources (RES). This transition is further bolstered by the widespread adoption of net-zero emissions pledges and a substantial advancement made in meeting the commitments of the Paris Agreement in 2015, with the aim of keeping the rise in global temperatures to below 2°C over pre-industrial levels [1]. Based on data provided by the Global Wind Energy Council (GWEC) [2], GWEC's study projects that in 2024, new installations will surpass the past record and reach 130 GW. Under present policy projections, an additional 791 GW of capacity is probably to be built in the next five years. Up until 2028, this comes to 158 GW of additional installations annually [2]. Though the amount of installed capacity for 2023 was the greatest in history, the expected compound annual growth rate for the next five years is rather close to 10% [2]. According to GWEC's projections, there will be a significant increase in new capacity from 2022 to 2026, as illustrated in **Figure 1a**. It is true that renewable energy sources (RES) have experienced significant growth in the last ten years. The global installed capacity for wind and solar power has increased by 6-fold and 40-fold, respectively [4]. Meanwhile, according to the International Energy Agency (IEA), the total installed capacity of wind power is expected to skyrocket by 11-fold to over 8000 GW by 2050 [3]. Wind power and solar photovoltaics emerge as the frontrunners, significantly boosting the proportion of RES in the overall global electricity production from 29% in 2020 to an impressive 90% by 2050. As illustrated in **Figure 1b**, wind power development is growing rapidly. In addition, the growth of RES has been driven by the decrease in costs, which have significantly dropped due to economies of scale, competitive supply chains, and technological advancements. The reduction in costs and greater implementation have created a positive feedback loop that keeps pushing prices lower. Additionally, government support in the form of tax incentives, subsidies, and funding for research and development has been crucial [1]. Transitioning to renewable energy sources is crucial for a sustainable energy future, but it comes with technical challenges due to the unique characteristics of RES compared to conventional generators [5]. Due to the natural synchronous inertia of their spinning mass, traditional energy generators, such as steam or hydro turbines, are connected to the power system grid in a manner that supports the stability of the network's frequency [6, 7]. However, electronic converters are employed to interface RES with the power grid [8]. This interface creates a decoupling effect, resulting in RES operating independently from the power system inertia. As a result, the turbines in the RES do not directly interact with the grid's frequency dynamics. This means alternative methods are needed to ensure grid stability when power inputs fluctuate [6, 7].

The impact of the RES being partially decoupled from the grid impedes the buffers provided by the conventional generators in the event of a contingency [7]. The effects are of two-fold: (a) It hampers the natural response of the system, causing it to slow down. (b) It deprives the controllers and system operator of sufficient time to make necessary adjustments, as systems with reduced system inertia have a significantly higher rate of change of frequency (RoCoF) [7]. The difference in characteristics, specifically

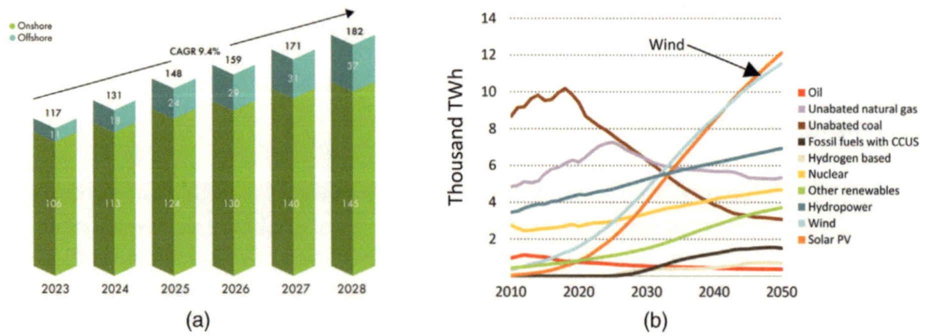

Figure 1.
a. Wind power outlook according to GWEC [2]. b. Global electricity outlook with respect to net-zero emissions by 2050 [3].

the partial decoupling of the renewable energy units from the grid, increases the grid's susceptibility to frequency deviations and a more rapid RoCoF, while also diminishing system inertia [8, 9]. Also, there are other characteristics that further exacerbate the issues of RESs, particularly wind plants, in power systems compared to conventional plants. One of these factors is the intermittency of renewable energy sources, which has a negative impact on power system operations. This variability makes it difficult to maintain system stability, especially regarding frequency stability and balancing power supply and demand. Additionally, wind turbines typically operate using a maximum power point tracking technique, which means they cannot generate extra power for low-frequency events. There have been events reported in literature where the inability to keep frequency deviations or maintain system RoCoF at a healthy level has led to cascaded blackouts, as experienced in Southern Australia and Great Britain in 2016 and 2019, respectively. In the case of Southern Australia, wind energy contributed the most to the grid's generation mix, around 48%. However, the UFLS (Under-Frequency Load Shedding) scheme was activated due to the wind farms' inability to withstand the deviations. This event has led to a Rate of Change of Frequency (RoCoF) value of 6.25 Hz/s, causing the system frequency to drop to 47 Hz within a time span of 0.4 seconds [7, 10].

However, it is essential to note that the reduced cost associated with renewable energy sources from production to installation presents viable prospects for delving deeper into the advantages of these resources in providing ancillary services, such as frequency regulation for ensuring the stability of the power system through tuning pitch control at a sub-optimal level when wind resources are abundant. This is crucial as frequency regulation is currently reliant solely on traditional generation methods, which are experiencing a rise in cost [11, 12]. According to J. Aho et al. [13], wind plants have not traditionally been obligated to offer grid regulation services, given that most modern wind turbines do not inherently deliver any of the grid regulation services that conventional generators offer. However, this is changing as nations with high levels of wind penetration in nations such as Spain, Ireland, and Denmark have prompted their transmission system operators to introduce fresh criteria for forthcoming wind plant installations to possess the ability to supply power tracking and frequency regulation services in times of abundant wind resources.

1.1 Contribution

The chapter introduces a supplementary control loop (secondary control scheme) to enable Wind Power Plants (WPPs) to participate in frequency regulation, which is a novel approach compared to existing methods. The originality of this research lies in developing and evaluating this control scheme using the IEEE 9-bus network model in DIgSILENT software. The study's findings demonstrate that WPPs can effectively contribute to secondary frequency control, potentially reducing reliance on fossil fuel-based generators for ancillary services. This is a significant contribution as it provides a new method for integrating renewable energy sources, particularly wind, into grid stability operations, which have traditionally been dominated by conventional generation sources.

2. Frequency control in the presence of wind power penetration

As mentioned earlier, rapid technological advancements have created a demand for new fast frequency response (FFR) service schemes to adapt to the evolving power system. These schemes work together with the three classical frequency control stages to prevent and restore frequency to its normal value after a contingency event, as shown in **Figure 2**. These services offer immediate active power to quickly restore system balance and prevent frequency from exceeding the under-frequency relay point in the power system during emergencies. Although the FFR and inertia services play similar roles in controlling system frequency and ensuring smooth frequency response during contingencies, they are delivered through different physical mechanisms within the system [15]. Various naming conventions are employed by different system operators worldwide when it comes to FFR services. For instance, National Grid (TSO, GB) is implementing a rapid reserve service. Similarly, EirGrid (TSO, Ireland) has put forward a new ancillary service product that needs to be activated within 2–3 seconds of a frequency event and maintained for a minimum of 8 seconds. In addition, ERCOT (TSO, Texas), the Australian energy market operator (AEMO), and Hydro-Quebec TSO have put forward a proposal for a fast frequency response ancillary service [12, 16].

Thus, it has become imperative for RES to play a more proactive role in modern power systems' frequency regulation, offering active power support in the event of a contingency. There are various approaches proposed to provide frequency support from RES, particularly wind plants, which is the focus of the study as shown in **Figure 3**.

Utilizing power converters with the maximum power point tracking (MPPT) technique allows for optimal power extraction from RES [15]. As mentioned earlier, these converters prevent direct interaction with the grid, resulting in no contribution to system inertia. In this section, we will explore the methods used to incorporate RES into the grid and the techniques used for frequency control, particularly pitch angle control is employed in this study as highlighted in **Figure 3**. With respect to wind plants, variable speed wind turbine generators (VSWTG) are widely favored. There are different types of wind turbine generators (WTGs) based on their speed, including fixed/semi-fixed-speed WTG (Type 1 and 2) and variable speed WTG (Type 3 and 4), when considering their response to inertia. The inertial response

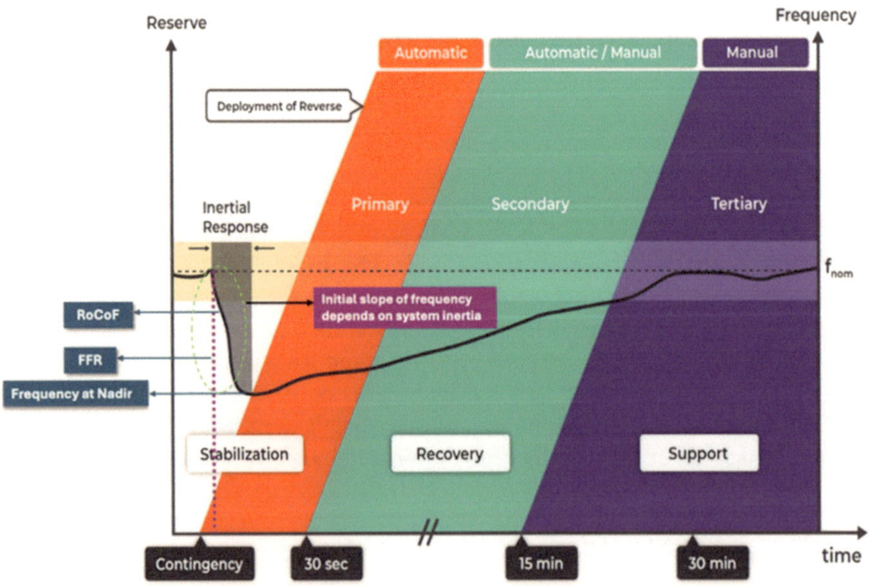

Figure 2.
Synchronized collaboration between frequency control stages after a contingency. Adapted from [14].

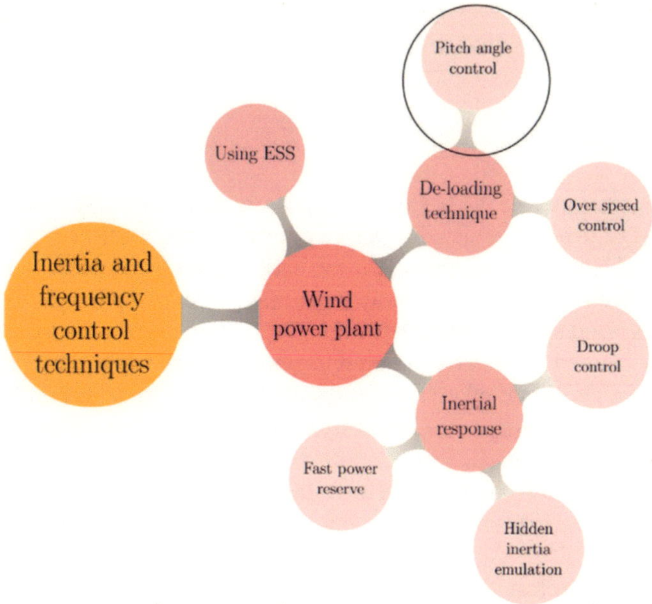

Figure 3.
Inertia and Frequency response techniques from wind plants. Adapted from [15].

of type 1 and type 2 WTGs is comparatively slower and lower when compared to synchronous generators and type 3 and 4 WTGs, which are depicted in **Figure 4**. Fixed-speed WTGs have a slightly smaller inertia constant due to a lower coupling of induction generator rotational speed to system frequency. Types 3 and 4 VSWTGs have minimal inertia due to the fast control of power electronic converters, which ensures a consistent output power regardless of fluctuations in grid frequency [16, 17].

With either of these turbines, they participate in frequency response employing either overproduction or de-loading techniques. It is imperative to note that the former scheme, due to mechanical stability constraints, can only be applied for a

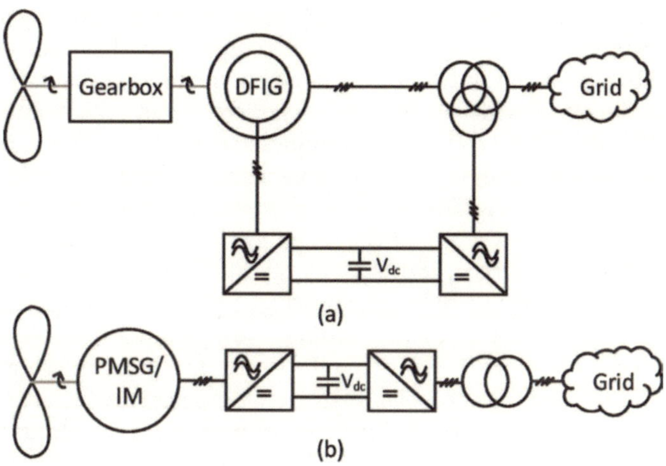

Figure 4.
Variable speed wind turbines a) type 3 b) type 4 [16].

short period, i.e. fast frequency response, however, the de-loading scheme can be applied for an extended period to support both fast frequency and primary reserve timeframes [15–17].

2.1 Overproduction scheme

Using this technique, VSWTG can participate in frequency response provision for a limited amount of time by generating active power beyond its rated output during high wind speed conditions and exceeding the available power during lower wind speed conditions. Typically, VSWTG can generate excess power of up to 0.2 pu for a duration of 10s when the wind speed is relatively high [16]. When the wind speed exceeds the rated limit, the blade pitch can be adjusted to generate more power in response to a drop in grid frequency. However, when the wind speed is below the rated limit, the turbine is set to maximize energy capture. This means that any extra power generated in the short term must come from the stored energy of the rotor, which results in a decrease in turbine rotor speed. There are three different approaches: i) droop control, ii) hidden inertia emulation, and iii) fast power reserve [15]. In the droop control approach, wind plants (WP) imitate the actions of a governor in a conventional SG to promptly react to any fluctuations in the system frequency. This indicates that the active power provided by the wind plants is adjusted in proportion to the frequency deviation. In the hidden inertia emulation approach, WP contributes to FFR by simulating the inertial response of traditional SGs. Nevertheless, the inertia response from wind turbine rotors is significantly smaller compared to that of SGs. The third approach involves the fast power reserve, which relies on the participation of WP to supply the kinetic energy stored in the rotating masses of the wind turbine. This is done in response to frequency deviation, similar to the hidden inertia approach. On the other hand, the fast power reserve approach stands out because it allows for the recovery of power during periods of underproduction. This recovery period is crucial for restoring both the power and the rotational speed of the wind turbine to their original values [18].

2.2 De-loading scheme

Addressing the energy recovery issue linked to overproduction and ensuring a sustained response that goes beyond the fast frequency response time window necessitates the de-loading of WTG. With its de-loaded operation, the headroom in WTG can be utilized to offer rapid frequency response and primary/secondary frequency support. The inclusion of wind turbines in the fast frequency reserve was initially discussed in [19], along with different techniques to ensure a quick emulated inertial response, ample primary reserve, and stable mechanical operation of the turbine. Among the various de-loading techniques proposed, **Figure 5(a)** illustrates a limitation in the maximum output power of the turbine at high wind speeds. This limitation is implemented to ensure an active power margin, but it is important to note that this reserve is not available below the rated speed. Another technique, shown in **Figure 5 (b)**, reduces the load on the turbine below its rated speed. This allows for the provision of reserve power by reducing the load at low wind speeds and increasing production at high wind speeds. Unfortunately, it is not feasible to offer a primary frequency reserve above rated speed since the turbine cannot handle prolonged overloading. As depicted in **Figure 5(c)**, the turbine has the capability to be de-loaded across its entire speed operating range, allowing for rapid and primary frequency responses, which is the approach employed in this study [16].

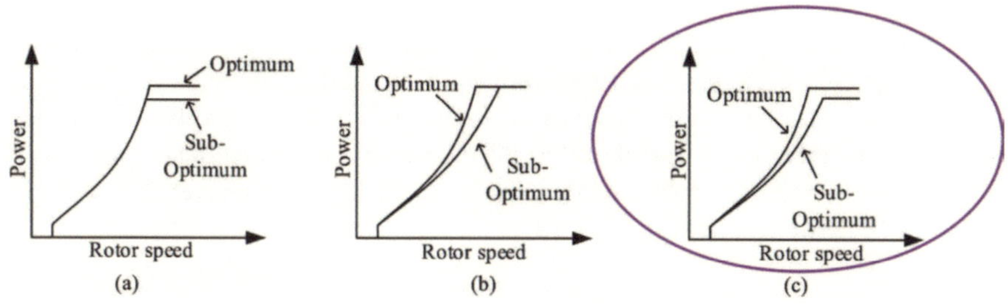

Figure 5.
Different WT de-loading characteristics a) de-loading over rated speed b) de-loading below rated speed c) de-loading overall speed [16].

In essence, VSWTG essentially has two options for running with the de-loading technique: (a) overspeed control and (b) pitch angle control. Maintaining the rotor speed at the maximum power point Ù $_{MPP}$, the pitch angle control consists of raising the pitch angle from β_0 *to* β_1 for a constant wind speed v, as depicted in **Figure 6a**. In this sense, $P_{del's}$ supplied power is less than the maximum aerodynamic power P_{MPP}. Consequently, a specific amount of active power reserve is ready to provide extra generation in the event of a frequency contingency. Maintaining the pitch angle β_0 at a constant wind speed V_w, the overspeed control moves the de-loaded power P_{del} toward the right of

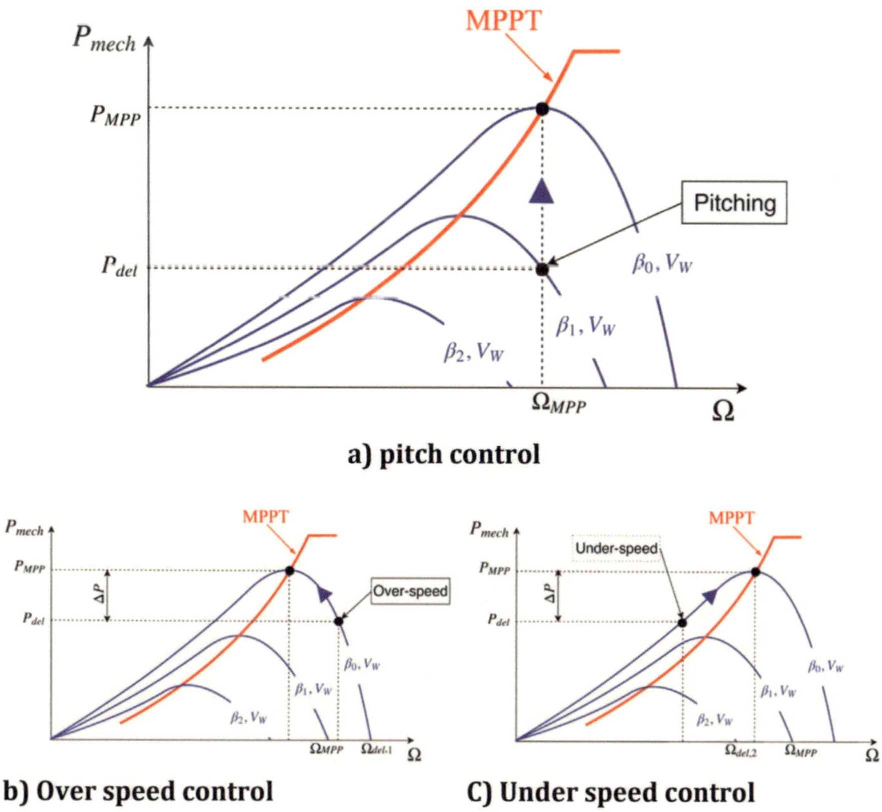

Figure 6.
De-loading techniques for Variable Speed WTG [15].

the maximum power P_{MPP}, as illustrated in **Figure 6b**. Reduced rotor speed from $\grave{U}_{del,1}$ to $\grave{U}_{MPP}$ releases kinetic energy to the system when frequency response is needed. A third possibility is shown in **Figure 6c**; in this scenario, the turbine will operate the rotor speed below rated rotor speed for MPPT operation. In this scenario, power is taken from the turbine to raise the rotor speed from $\grave{U}_{del,2}$ to $\grave{U}_{MPP}$. However, this approach is regarded as a "hazardous strategy" [15, 18].

3. Wind turbine generator control models and secondary frequency control scheme

The structural framework of wind energy conversion systems classified as Types 3 and 4 comprises seven and four distinct modules, respectively [20]. This modular framework was instituted to depict diverse plant configurations and enhance the capability to update individual model constituents as technological advancements occur and/or additional data is procured.

The modules that characterize a generic wind energy conversion system of Type 3 include Converter, Electrical Control, Aerodynamic Conversion, Pitch Controller, Drive-Train, Torque Controller, and Plant-level Controller. The modules necessary for representing a Type 4 Wind Turbine Generator (WTG) power plant are a subset of those essential for modeling a Type 3 WTG facility. Modules whose nomenclature commences with the letters "r" and "e" (denoting renewable energy) are components that may also serve to depict photovoltaic energy generation systems, see **Figure 7** [21]. In this revision of the models, each nomenclature of the components is appended with "_a" to signify the first revision of the module, as indicated in **Figure 7** [21].

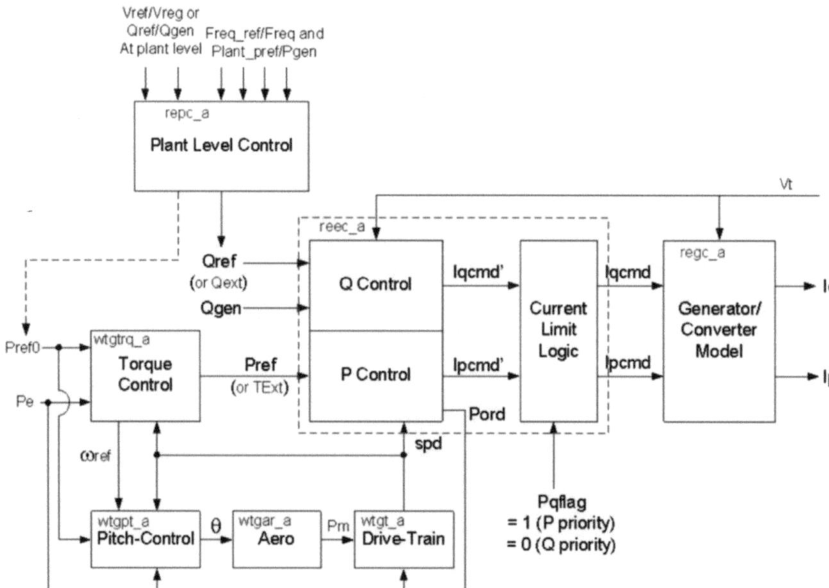

Figure 7.
A feedback loop system for wind power plant-level control strategy [21].

This research primarily examines the type 3 model, which is driven by the advantages associated with the individual control model. Type 3 exhibits a greater variety of individual models as indicated in **Figure 7**. The literature indicates that frequency control utilizing wind turbine generators can be implemented by modifying the pitch control model and incorporating an additional control loop [22].

Moreover, it is significant to highlight that, at present, the predominant technologies installed for new wind energy facilities, both in the United States and internationally, are the type 3 and type 4 units. While many type 1 and 2 units remain operational globally, their applicability for various services is considerably restricted. All principal equipment manufacturers provide one or both technologies (types 3 and 4) [22].

3.1 Aerodynamic mathematical model

The pitch angle is the main input that the aerodynamic model takes into account. The aerodynamic model calculates mechanical output power using pitch angle and rotor speed. Eq. 1 can be used to determine the mechanical output power generated by wind turbine generators [23].

$$P_m = \frac{1}{2} C_p(\lambda,\beta)\rho A v^3 \tag{1}$$

Where A is the blade-swept area, v is the wind speed, λ is the tip speed ratio, commonly known as lambda, β is the blade pitch angle, ρ is the air density, P_m is the mechanical output power, and C_p is the power coefficient.

The swept area is determined by Eq. 2. Here, R denotes the rotor radius.

$$A = \pi \times R^2 \tag{2}$$

Eq. 3 is used to determine the tip speed, where ω_T represents the hub or turbine's rotating speed.

$$\lambda = \frac{\omega_T R}{v} \tag{3}$$

The correlation between mechanical power and blade pitch angle can be mathematically represented, as indicated in eq. 4.

$$P_{mech} = P_{mech0} - K_{areo}\beta(\beta - \beta_0) \tag{4}$$

Where P_{mech0} is the initial mechanical power, β_0 is the initial hub or blade pitch angle.

3.2 Wind turbine drive-train mathematical model

The aerodynamic model's output drives the turbine and generator as one unit. Eq. 5 provides the mathematical representation of the wind turbine's speed ω_t. H is the total inertial constants of the generator and turbine, along with D, the shaft's damping factor, and ω_0, the turbine's initial speed, $\Delta\omega_t$ is the speed deviation of the turbine [22].

$$\omega_t = \frac{1}{2H(\omega_0+\Delta\omega_t)}(P_{mech}\text{-}P_e\text{-}D\Delta\omega_t) \tag{5}$$

From the drive-train model, the change in rotor speed from its reference point serves as an input for both pitch and torque control.

3.3 Pitch control model

The amount of mechanical energy produced by the wind turbine generator is determined by the pitch angle control. When wind energy is generated and the required amount of power is not met by the grid, the pitch angle controller is activated. This controller makes certain that the wind turbine's energy is extracted as efficiently as possible by varying the pitch of the blades. Two PI controllers assess the power and speed disparities to manage the blade pitch's dynamic reaction [22]. The modified pitch control model is illustrated in **Figure 8**, with its corresponding mathematical models from Eqs. 6-12, representing the model.

The mathematical representation of the pitch compensation mechanism is delineated in Eq. 6.

$$\dot{\theta} = \frac{1}{T_{pi}}(\theta_{pc} + \theta_{pco} - \theta) \tag{6}$$

where

$$\theta_{pc} = K_{pw}\left[\omega_t - \omega_{ref} + K_{cc}\left(P_{ord} - P_{ref} + dPWTG\right)\right] + s_0 \tag{7}$$

and

$$s_0 = \int_{T_{min}}^{T_{max}}\left[\omega_t - \omega_{ref} + K_{cc}\left(P_{ord} - P_{ref} + dPWTG\right)\right]dx \tag{8}$$

The signal s_0 refers to the speed PI control.

While

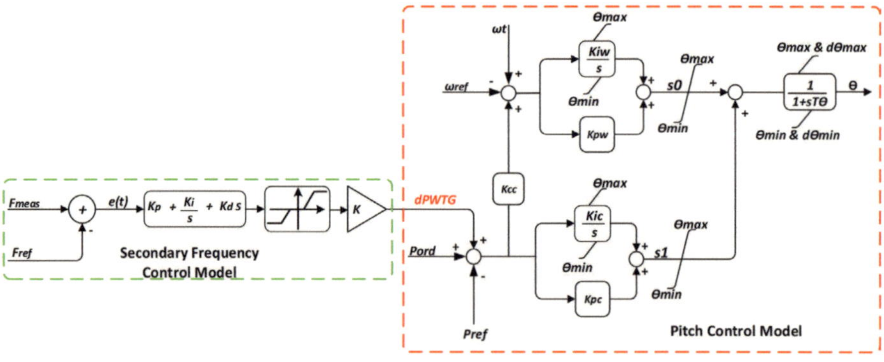

Figure 8.
Modified Pitch control with frequency control loop.

$$\theta_{pco} = K_{pc}\left(P_{ord} - P_{ref} + dPWTG\right) + S_1 \tag{9}$$

where

$$S_1 = \int_{T_{min}}^{T_{max}} \left(P_{ord} - P_{ref} + dPWTG\right) dx \tag{10}$$

The signal S_1 denotes PI control of active power.

Figure 8 depicts the supplementary control loop designed to facilitate the participation of wind turbine generators in frequency response, with emphasis placed on the area delineated in green. Within this control framework, f_{meas} represents the actual frequency measurement obtained from the system. f_{ref} denotes the desired reference frequency intended for the system. The variation in frequency is calculated as per Eq. 11, $e(t)$, which is the difference between the reference frequency and the measured frequency.

$$e(t) = f_{meas} - f_{ref} \tag{11}$$

This is the error signal input to the controller. After the PID controller, a limiter restricts the control signal to a specific range. This is used to prevent excessive pitch angles, which might damage the turbine or lead to instability. The limiter output controls the pitch actuator, $K_{actuator}$, which adjusts the blade pitch to regulate the rotor speed and, consequently, control the frequency of the generated power. Eq. 12 represents the overall secondary control scheme.

$$dPWTG = \left(\left(K_p e(t) + K_i \int_0^t e(t)dt + K_d \frac{de(t)}{dt}\right) \cdot K_{actuator}\right) \tag{12}$$

3.4 Torque control model

To maintain the initial electrical torque, the torque control is in place. Two methods can be employed to ascertain the torque reference. Using maximum power tracking is the first technique. For the purpose of achieving the most torque possible, this approach will attempt to bring the turbine closer to the rotor speed. An alternative method is to generate the electrical torque using a plant controller, which establishes the active power reference [22].

4. Ziegler-Nichols closed-loop tuning method

This chapter employed the Ziegler-Nichols second methodology to optimize the PID parameters. The Ziegler-Nichols closed-loop tuning technique facilitates the utilization of the ultimate gain value, Ku, along with the ultimate oscillation period, Pu, to determine Kc. This approach offers a straightforward framework for configuring PID controllers and can be fine-tuned to yield more precise estimates of the control settings. The tuning constants Kc, Ti, and Td can be derived within a feedback control system. It is essential to recognize that the Ziegler-Nichols closed-loop tuning technique is restricted to processes incapable of operating in an open-loop configuration [24].

The ultimate gain value, Ku, is assessed by identifying the proportional-only gain that induces the control loop to oscillate perpetually in a steady state. This involves setting both the integral and derivative controller gains to zero, effectively isolating the influence of the proportional gain. This process assesses the stability of the Kc value to enhance it for the controller's performance. The ultimate period is another significant variable associated with this proportional-only control tuning method (Pu). The ultimate period denotes the duration required to complete one full oscillation while the system maintains a steady state. These two parameters, Ku and Pu, constitute the basis for computing the loop-tuning constants of the controller (P, PI, or PID). To ascertain the values of these parameters and derive the tuning constants, one should follow the procedure outlined in the flow chart in **Figure 9**.

The proposed control strategy, illustrated in **Figure 10**, is assessed using an adapted IEEE 9-bus test system incorporating wind power plants. This well-established test system is frequently utilized for stability assessments, particularly focusing on frequency stability [25]. The IEEE 9-bus standard consists of nine busses, three generators, three loads, and three transformers. Busses 2 and 3 function as PV busses due to their attached generators, while busses 4 through 9 operate as PQ busses. The reference bus or slack bus is designated as bus 1. Busses 1, 2, and 3 are interconnected with a generator via three step-up transformers. Loads are associated with busses 5, 6, and 8, respectively.

Table 1 delineates the information pertaining to the three generators, three loads within the network, the classification of busses, as well as the voltage levels at which these components are interconnected. The specifications of the transformer and line are presented in **Tables 2** and **3**, correspondingly.

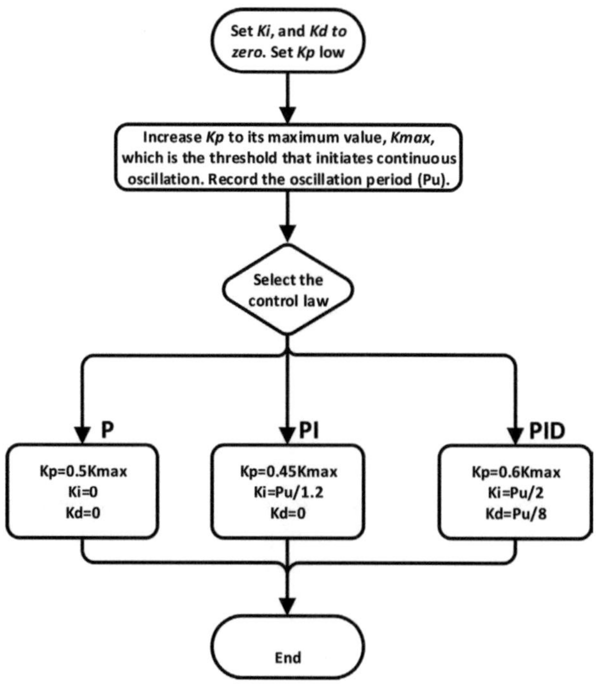

Figure 9.
Ziegler-Nichols method for tuning P, PI, and PID controllers [24].

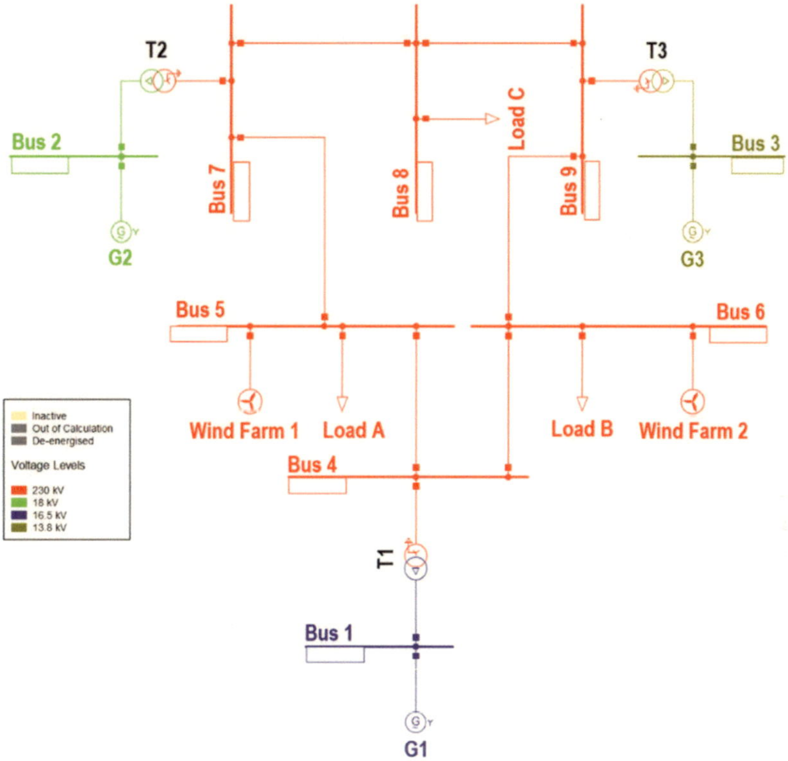

Figure 10.
Modified IEEE 9-Bus Network Integrated with Wind Power Plants.

In this configuration, wind power plants are positioned at busses 5 and 6. As implemented here, the strategic placement of renewable energy sources in proximity to the load is essential for reducing power losses by minimizing transmission needs and facilitating more effective power delivery.

Bus No.	Type	Generation		Load		Bus Voltage	
		MW	Mvar	MW	Mvar	(pu)	kV
1	Slack	71.6	27	0	0	1.04	16.5
2	PV	163	6.7	0	0	1.025	18.0
3	PV	85	−10.9	0	0	1.025	13.8
4	PQ	0	0	0	0	1.026	230
5	PQ	0	0	125	50	0.996	230
6	PQ	0	0	90	30	1.013	230
7	PQ	0	0	0	0	1.026	230
8	PQ	0	0	100	35	1.016	230
9	PQ	0	0	0	0	1.032	230

Table 1.
IEEE 9-bus Generator and Load data [26].

	X (pu)	MVA	Voltage ratio
T1	0.0576	77	16.5/230 kV
T2	0.0625	163	18/230 kV
T3	0.0586	86	13.8/230 kV

Table 2.
IEEE 9-bus Transformer data [27].

Line from	Line to	Resistance (R) (pu)	Reactance (x) (pu)	Susceptance (B) (pu)
4	5	0.01	0.085	0.176
4	6	0.017	0.092	0.158
6	9	0.039	0.17	0.358
5	7	0.032	0.161	0.306
9	3	0	0.0586	0
7	2	0	0.0625	0
9	8	0.0119	0.1008	0.209
7	8	0.0085	0.072	0.149

Table 3.
IEEE 9-bus Line data [26].

5. Simulations and discussion

As the electrical power infrastructure progresses, the reliance on fossil fuel-based generators for secondary control mechanisms is anticipated to diminish. As previously highlighted in the introductory section, the financial burden associated with fossil fuel generators engaged in secondary control is escalating. Although wind turbine generators have not been employed in practical applications thus far, they exhibit potential for favorable results. Several scholarly investigations have suggested the feasibility of integrating wind turbine generators into secondary control; however, the methodology for achieving this integration remains ambiguous. The existing body of literature proposes that techniques such as de-loading and overproduction may be applicable for this purpose. Nevertheless, the specific elements of the control framework that are influenced by these techniques remain uncertain. To validate the assertions presented in the literature and facilitate the implementation of wind turbine generators, the control strategy delineated in Section 3.3 is rigorously assessed across a range of contingencies. This strategy represents the sole approach enabling wind turbine generators to engage in secondary control, thereby eliminating the dependence on costly fossil fuel-based conventional generators, which are detrimental to the environment. Consequently, wind turbine generators emerge as the exclusive technology within this research capable of restoring the system frequency to its nominal value following a disturbance. The two contingency scenarios employed to evaluate the proficiency of a wind turbine generator in engaging in secondary frequency regulation encompass an increase in load demand and the generator trip event at varying levels of wind energy penetration. **Table 4** delineates the data associated

	Number of WTGs		Total Capacity (2 MW)		Scheduled Power	
	Farm 1	Farm 2	Farm 1	Farm 2	Farm 1	Farm 2
WTG @ 30%	30	30	60 MW	60 MW	48 MW	48 MW
WTG @ 50%	50	50	100 MW	100 MW	80 MW	80 MW
WTG @ 60%	60	60	120 MW	120 MW	96 MW	96 MW

Table 4.
Wind Turbine Generator (Type 3) Capacity at Different penetration levels.

with the wind turbine generator, including its capacity across different penetration levels. Furthermore, it incorporates the quantity of power allocated at these diverse penetration levels.

5.1 Load demand increase

The effectiveness of the proposed control strategy is assessed through a contingency that involves a 10% increase in load demand initiated at t = 5 s. This situation is examined across various levels of wind power plant integration. The diagram in **Figure 11** illustrates the frequency response under two scenarios: (a) without secondary frequency control and (b) with secondary frequency control. In scenario (a), after the load increase, the system undergoes a notable initial drop in frequency, referred to as the frequency nadir. Although primary control measures assist in partially recovering the frequency, the system ultimately stabilizes at a frequency below the nominal value, resulting in a steady-state deviation. This deviation is particularly pronounced in systems with greater wind penetration, where the reduced inertia of wind turbines causes a more profound nadir and a slower recovery. However, scenario (b) showcases the benefits of secondary frequency control in effectively correcting frequency deviations and enhancing the frequency nadir. While the initial dip still takes place, the system is able to recover more efficiently and gradually returns to its nominal frequency of 50 Hz, with an improved frequency nadir. This control mechanism is especially vital in scenarios with high wind penetration, as it mitigates the instability risks linked to diminished system inertia. Therefore, secondary frequency control is crucial for sustaining grid stability, ensuring that the system frequency is restored and maintained at its designated operational level.

The two graphs in **Figure 12** depict the impact of wind power contribution during a load increase at various wind penetration levels, emphasizing the role of inertial

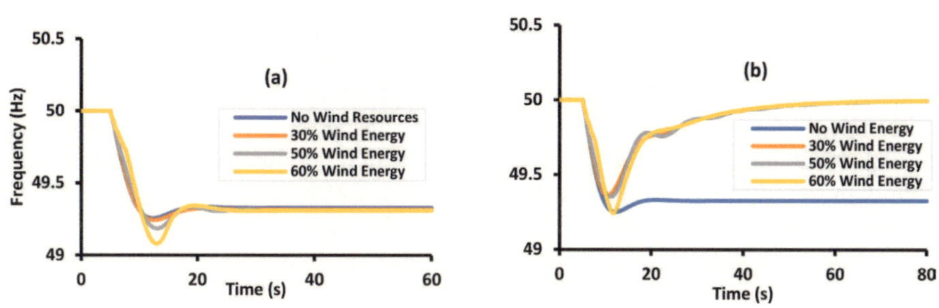

Figure 11.
System Frequency Response Following a Load Increase at Different Wind Power Penetration Levels with Inertial Control: (a) Without Secondary Frequency Control, (b) With Secondary Frequency Control.

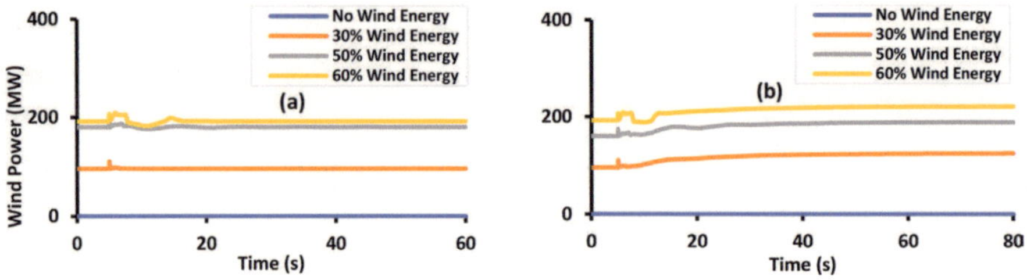

Figure 12.
Wind Power Contribution Following a Load Increase at Different Wind Power Penetration Levels with Inertial Control: (a) Without Secondary Frequency Control, (b) With Secondary Frequency Control.

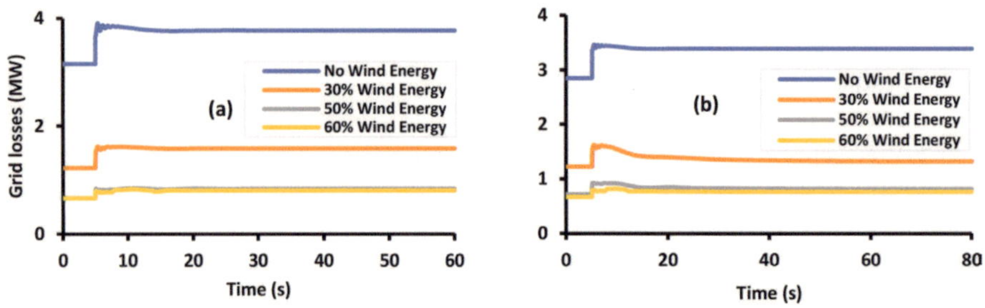

Figure 13.
Power losses Following a Load Increase at Different Wind Power Penetration Levels with Inertial Control: (a) Without Secondary Frequency Control, (b) With Secondary Frequency Control.

control with and without secondary frequency control. In the absence of secondary frequency control, the wind power response to load increases is limited. At 30% penetration, there is a minor initial power increase that quickly stabilizes, indicating a weak response. As penetration rises to 50% and 60%, the response becomes more notable, especially at 60%, where a significant increase occurs but stabilizes at a higher level than at lower penetrations. Nevertheless, the overall reaction remains subdued, suggesting that inertial control alone is inadequate for optimizing wind power utilization during abrupt load variations. Conversely, the implementation of secondary frequency control significantly enhances wind power contribution across all penetration levels. At 30% penetration, the response improves slightly compared to the absence of this control, but it remains modest. Notably, there is a substantial enhancement at 50% and 60% penetration, with the 60% level showing a rapid increase followed by stabilization at a higher output, illustrating a robust response to load changes. This indicates that secondary frequency control effectively augments the system's capacity to leverage wind power, particularly at elevated penetration levels.

The two graphs in **Figure 13** illustrate the effects of secondary frequency control on power losses during load increases at varying wind power levels. The absence of secondary frequency control results in an increase in power losses, especially with less wind energy integration. For example, power losses remained around 3.5 MW without wind energy. In contrast, at 60% wind penetration, they decline to about 1.5 MW, indicating that increased wind penetration mitigates power losses, albeit insufficiently without secondary frequency control.

Conversely, the implementation of secondary frequency control markedly reduces power losses across all penetration levels. In the absence of wind energy, losses were approximately 3.2 MW. The most significant reductions occur at elevated

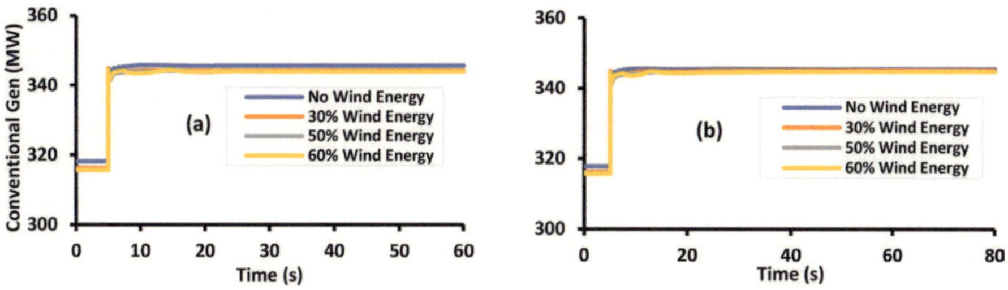

Figure 14.
Power contribution from conventional generators in response to Load Increase at Different Wind Power Penetration Levels with Inertial Control: (a) Without Secondary Frequency Control, (b) With Secondary Frequency Control.

wind penetration levels, with losses dropping to just above 1 MW at 60% penetration, underscoring the efficacy of secondary frequency control in improving grid efficiency. The analysis illustrates that secondary frequency control enhances the advantages of increased wind energy penetration and reduces power losses, thereby promoting a more efficient power system.

Figure 14 illustrates the power output from conventional generators in response to load increases under varying wind power penetration levels, contrasting scenarios with and without secondary frequency control. In the absence of secondary frequency control, conventional generators exhibit a rapid power increase post-load surge, reaching approximately 345 MW and stabilizing thereafter, with consistent performance across wind penetration levels. This uniform response implies a substantial reliance on conventional generators, regardless of wind integration. With secondary frequency control, conventional generators display a marginally more efficient response, characterized by a similar initial power spike but a quicker and slightly lower stabilization, particularly notable at elevated wind penetration levels. Specifically, at 60% wind penetration, the contribution of conventional generators stabilizes at lower levels compared to the scenario without secondary control because more power is delivered by wind power plants, indicating enhanced power distribution optimization.

The major distinction resides in the efficiency of conventional generators' responses. Without secondary frequency control, reliance on conventional sources remains significant and uniform across wind levels. Conversely, secondary frequency control facilitates improved efficiency, diminishing the dependency on conventional generation as wind penetration increases, thereby underscoring its importance in augmenting the efficiency and stability of the power system through better wind energy integration.

5.2 Generator trip

The performance of the proposed control strategy was additionally assessed across various contingencies, including the generator trip at multiple wind energy penetration levels. The contingency was also applied at t = 5 s. The diagrams in **Figure 15** illustrate the system frequency response following the trip of Generator 3 at differing wind power levels with inertial control. The left diagram (a) specifies the response without secondary control, whereas the right diagram (b) incorporates secondary control. In diagram (a), the frequency experiences a sharp decline post-trip, with more pronounced drops at elevated wind penetration levels, signifying diminished system inertia. Following the decline, the frequency stabilizes at a reduced level, with increased wind resulting in a lower steady-state frequency. Diagram (b) depicts the frequency response with secondary control, where an initial drop occurs, but recovery is markedly improved. Secondary control mitigates the frequency dip and enables recovery closer to the nominal frequency, stabilizing at a higher level even at 60% wind energy. In conclusion, secondary frequency control significantly enhances system resilience against frequency deviations, facilitating recovery and stabilization in high-wind scenarios.

The graphs in **Figure 16** compare wind power contributions during a generator trip with and without secondary frequency control at different penetration levels. The left graph (a) without secondary control shows significant fluctuations in wind power contributions, particularly at higher penetration levels like 60%, where there is a sharp rise followed by a decrease. In contrast, (b) with secondary control displays smoother wind power contributions, especially at higher levels, indicating better stability due to the control measures. Overall, it is observed that the secondary frequency control reduces sharp fluctuations and enhances system stability, especially with higher wind energy penetration.

The two graphs illustrated in **Figure 17** compare power losses after a generator trip at various wind power levels, with and without secondary frequency control.

In **Figure 17(a)**, without secondary frequency control, power losses peak sharply after a generator trip and stabilize at 4.5 MW, with lower initial losses as wind penetration increases, but some fluctuation remains. In **Figure 17(b)**, with secondary frequency control, the response is more stable, with less pronounced initial peaks

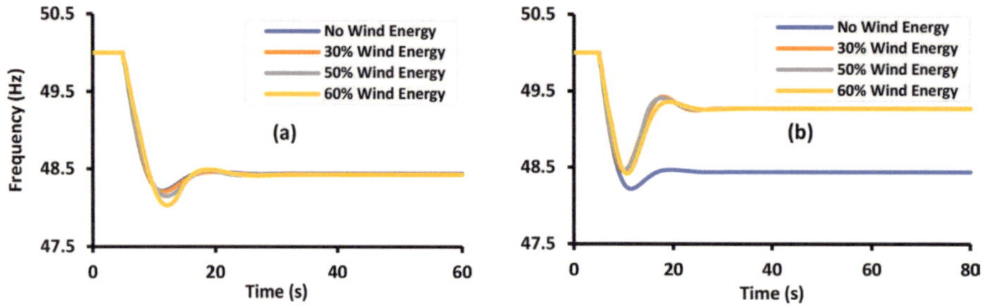

Figure 15.
System Frequency Response Following Gen 3 Trip at Different Wind Power Penetration Levels with Inertial Control: (a) Without Secondary Frequency Control, (b) With Secondary Frequency Control.

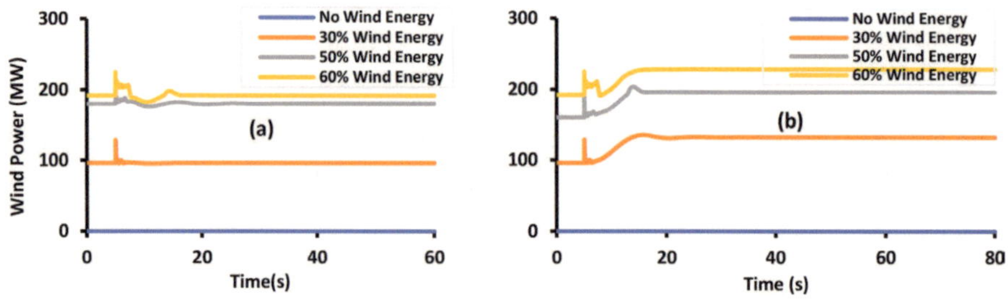

Figure 16.
Wind power contribution in response to Gen 3 Trip at Different Wind Power Penetration Levels with Inertial Control: (a) Without Secondary Frequency Control, (b) With Secondary Frequency Control.

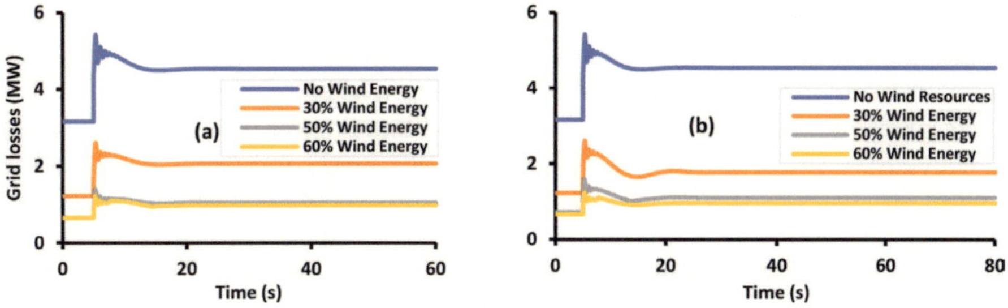

Figure 17.
Power losses Following Gen 3 Trip at Different Wind Power Penetration Levels with Inertial Control: (a) Without Secondary Frequency Control, (b) With Secondary Frequency Control.

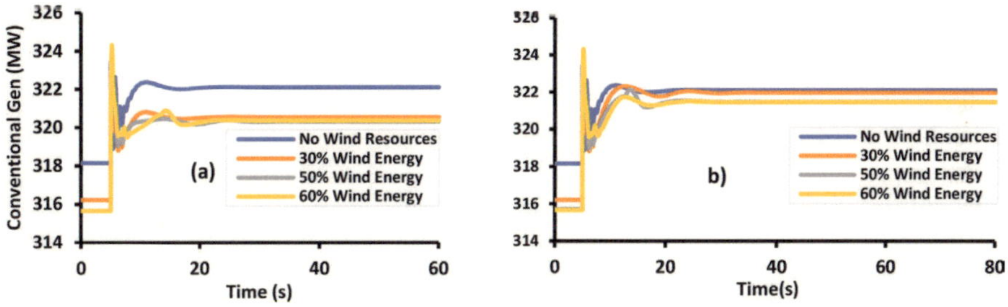

Figure 18.
Active Power from Conventional Generator Following Gen 3 Trip at Different Wind Power Penetration Levels with Inertial Control: (a) Without Secondary Frequency Control, (b) With Secondary Frequency Control.

and quicker stabilization, showing lower losses at higher wind penetration levels with minimal fluctuations. The secondary frequency control significantly minimizes power losses and facilitates faster stabilization during disturbances, enhancing the integration of wind energy.

The two graphs in **Figure 18** illustrate how a conventional generator's active power reacts to a trip at varying levels of wind power, comparing scenarios with and without secondary frequency control. In **Figure 18(a)**, the absence of secondary frequency control leads to significant fluctuations in active power after the generator trip, particularly at higher wind penetration levels, with a sharp drop and oscillations before stabilizing around 318 MW. At both 30% and 50% wind penetration levels, there are also noticeable variations from the steady-state value. **Figure 18(b)** demonstrates that with secondary frequency control, the active power response from the conventional generator is more stable, showing fewer fluctuations across all wind penetration levels, quickly stabilizing at around 318 MW even at 60% wind penetration. The secondary frequency control greatly enhances the stability of the conventional generator's active power output after a trip, especially at higher wind penetration levels, by reducing power fluctuations and facilitating quicker recovery, thereby improving overall grid stability.

6. Conclusions

This chapter presents an exhaustive examination of the complexities and methodologies linked to the incorporation of wind energy into contemporary power systems, with particular emphasis on the preservation of frequency stability. It commences by delineating the escalating global dependence on renewable energy sources, especially wind energy, and the ensuing difficulties engendered by the disconnection of renewable energy from the dynamics of grid frequency. The chapter investigates a variety of frequency response approaches, encompassing both overproduction and de-loading strategies, which facilitate the effective contribution of wind turbines to frequency regulation. Additionally, it introduces advanced control models and secondary frequency control mechanisms designed to alleviate the challenges introduced by the diminished inertia of wind power generation facilities. Through comprehensive case studies and simulations, the chapter illustrates the efficacy of these methodologies in sustaining grid stability across diverse scenarios, such as increases in load and generator outages. The chapter underscores the critical significance of innovative control mechanisms and technological progress in guaranteeing the dependable integration of wind energy into power systems.

Conflict of interest

The authors declare no conflict of interest.

Author details

Mkhutazi Mditshwa*, Azeez Olasoji, Komla Folly and David Oyedokun
University of Cape Town, Cape Town, South Africa

*Address all correspondence to: mdtmkh001@myuct.ac.za

References

[1] The Paris Agreement | UNFCCC. [Online]. Available from: https://unfccc.int/process-and-meetings/the-paris-agreement [Accessed: September 03, 2024]

[2] Global Wind Energy Council (GWEC). Global Wind Report 2024. [Online]. Available from: https://gwec.net/global-wind-report-2024/ [Accessed: August 28, 2024]

[3] International Energy Agency (IEA). Net Zero by 2050 - a Roadmap for the Global Energy Sector. [Online]. 2021. Available from: https://www.iea.org/reports/net-zero-by-2050 [Accessed: September 03, 2024]

[4] Tuo M, Li X. Security-constrained unit commitment considering locational frequency stability in low-inertia power grids. IEEE Transactions on Power Systems. 2023;**38**(5):4134-4147. DOI: 10.1109/TPWRS.2022.3215915

[5] Okafor CE, Folly KA. Optimal placement of BESS in a power system network for frequency support during contingency. Energy Reports. 2023;**10**:3681-3695. DOI: 10.1016/j.egyr.2023.10.017

[6] Tielens P, Van Hertem D. The relevance of inertia in power systems. Renewable and Sustainable Energy Reviews. 2016;**55**:999-1009. DOI: 10.1016/j.rser.2015.11.016

[7] Olasoji AO, Oyedokun DTO. Framework for ancillary services design for low inertia power systems. In: 2023 31st Southern African Universities Power Engineering Conference (SAUPEC). Johannesburg, South Africa: IEEE; 2023. pp. 1-6. DOI: 10.1109/SAUPEC57889.2023.10057797

[8] Fang J, Li H, Tang Y, Blaabjerg F. On the inertia of future more-electronics power systems. The IEEE Journal of Emerging and Selected Topics in Power Electronics. 2019;**7**(4):2130-2146. DOI: 10.1109/JESTPE.2018.2877766

[9] Milano F, Dörfler F, Hug G, Hill DJ, Verbič G. Foundations and challenges of low-inertia systems (invited paper). In: 2018 Power Systems Computation Conference (PSCC). Dublin, Ireland: IEEE; 2018. pp. 1-25. DOI: 10.23919/PSCC.2018.8450880

[10] Li L, Zhu D, Zou X, Hu J, Kang Y, Guerrero JM. Review of frequency regulation requirements for wind power plants in international grid codes. Renewable and Sustainable Energy Reviews. 2023;**187**:113731. DOI: 10.1016/j.rser.2023.113731

[11] Obaid ZA, Cipcigan LM, Abrahim L, Muhssin MT. Frequency control of future power systems: Reviewing and evaluating challenges and new control methods. Journal of Modern Power Systems and Clean Energy. 2019;**7**(1):9-25. DOI: 10.1007/s40565-018-0441-1

[12] Badesa L, Teng F, Strbac G. Simultaneous scheduling of multiple frequency Services in Stochastic Unit Commitment. IEEE Transactions on Power Systems. 2019;**34**(5):3858-3868. DOI: 10.1109/TPWRS.2019.2905037

[13] Aho J et al. A tutorial of wind turbine control for supporting grid frequency through active power control. In: 2012 American Control Conference (ACC). Montreal, QC, Canada: IEEE;

2012. pp. 3120-3131. DOI: 10.1109/ACC.2012.6315180

[14] Shazon MNH, Nahid-Al-Masood, Jawad A. Frequency control challenges and potential countermeasures in future low-inertia power systems: A review. Energy Reports. 2022;**8**:6191-6219. DOI: 10.1016/j.egyr.2022.04.063

[15] Fernández-Guillamón A, Gómez-Lázaro E, Muljadi E, Molina-García Á. Power systems with high renewable energy sources: A review of inertia and frequency control strategies over time. Renewable and Sustainable Energy Reviews. 2019;**115**:109369. DOI: 10.1016/j.rser.2019.109369

[16] Karbouj H, Rather ZH, Flynn D, Qazi HW. Non-synchronous fast frequency reserves in renewable energy integrated power systems: A critical review. International Journal of Electrical Power & Energy Systems. 2019;**106**:488-501. DOI: 10.1016/j.ijepes.2018.09.046

[17] Aziz A, Than Oo A, Stojcevski A. Frequency regulation capabilities in wind power plant. Sustainable Energy Technologies and Assessments. 2018;**26**:47-76. DOI: 10.1016/j.seta.2017.10.002

[18] Alsharif H, Jalili M, Hasan KN. Fast frequency response services in low inertia power systems—A review. Energy Reports. 2023;**9**:228-237. DOI: 10.1016/j.egyr.2023.05.193

[19] Ekanayake J, Holdsworth L, Jenkins N. Control of DFIG wind turbines. Power Engineer. Feb 2003;**17**(1):28-32(4)

[20] Sanchez-Gasca JJ. Generic wind turbine generator models for WECC - a second status report. In: 2015 IEEE Power & Energy Society General Meeting. Denver, CO, USA: IEEE; 2015. pp. 1-5. DOI: 10.1109/PESGM.2015.7285645

[21] Pourbeik P et al. Generic stability models for type 3 & 4 wind turbine generators for WECC. In: 2013 IEEE Power & Energy Society General Meeting. Vancouver, BC, Canada: IEEE; 2013. pp. 1-5. DOI: 10.1109/PESMG.2013.6672398

[22] Motta RT, Dotta D, Wilches-Bernal F. Development and assessment of second generation WTG models in an open source platform. Energy Systems. 2021;**12**(3):657-693. DOI: 10.1007/s12667-019-00337-z

[23] Boyle J, Littler T, Foley A. Review of frequency stability services for grid balancing with wind generation. The Journal of Engineering. 2018;**2018**(15):1061-1065. DOI: 10.1049/joe.2018.0276

[24] Ellis G. Chapter 6 - four types of controllers. In: Ellis G, editor. Control System Design Guide. Fourth ed. Boston: Butterworth-Heinemann; 2012. pp. 97-119. DOI: 10.1016/B978-0-12-385920-4.00006-0

[25] Peyghami S, Davari P, Fotuhi-Firuzabad M, Blaabjerg F. Standard test Systems for Modern Power System Analysis: An overview. IEEE Industrial Electronics Magazine. 2019;**13**(4):86-105. DOI: 10.1109/MIE.2019.2942376

[26] Channi HK, Sandhu R, Kaur H, Singh P, Gaba S, Kaur J. An approach for load flow and load dispatch analysis of IEEE-9 bus system. In: 2023 2nd International Conference on Ambient Intelligence in Health Care (ICAIHC). Bhubaneswar, India: IEEE; 2023. pp. 1-6. DOI: 10.1109/ICAIHC59020.2023.10431399

[27] Asija D, Choudekar P, Soni KM, Sinha SK. Power flow study and contingency status of WSCC 9 bus test system using MATLAB. In: 2015 International Conference on Recent Developments in Control, Automation and Power Engineering (RDCAPE). Noida, India: IEEE; 2015. pp. 338-342. DOI: 10.1109/RDCAPE.2015.7281420

Chapter 2

An Overview of Analysis and Adaptive Pitch Control of an Offshore Floating Multi-Wind-Turbine Platform

Srikanth Bashetty and Selahattin Ozcelik

Abstract

This chapter outlines the findings of research focused on the creation of a new semi-submersible Offshore Floating Multi-Wind-Turbine Platform (OFFWIND). The primary concerns addressed involve modeling the aerodynamic forces caused by wind speed on the turbines, as well as the hydrodynamic forces resulting from wind and wave interactions on the entire system. To simulate the rotation of the rotors on the five wind turbines placed on the platform, a multiple-moving reference frame combined with a sliding mesh technique is employed. The evaluation of pressure, velocity, and turbulence intensity contours of the OFFWIND is conducted, particularly concentrating on the downstream turbine which operates in the partial wake of the turbines located upstream. The aerodynamic forces acting on the wind turbines are assessed to analyze their aerodynamic efficiency. The determined aerodynamic loads are integrated with the hydrodynamic forces to assess the platform's movements in response to wind and wave conditions. An adaptive control algorithm has been developed for blade pitch control. The performance of the adaptive control is evaluated against a standard Proportional-Integral (PI) controller. Simulation results demonstrate that the adaptive controller significantly improves rotor speed stability and reduces power output fluctuations under varying operational conditions.

Keywords: offshore wind turbine, modeling, aerodynamics, hydrodynamics, adaptive control

1. Introduction

Recent advancements in technology and supportive policies worldwide have fostered growth in the offshore wind industry. Offshore floating wind turbines hold great potential for tapping into vast wind energy resources available on the continental shelf, benefiting coastal communities globally. For water depths exceeding 50 meters, these floating platforms can harvest wind energy while minimizing visual and environmental concerns. The idea of installing multiple turbines on a single floating platform is gaining traction, as it offers the potential to reduce installation and mooring expenses while improving stability [1].

Advancements in technology have positioned this multi-turbine approach as a competitive solution for generating substantial wind energy [2–10]. Numerous studies have explored the dynamics of single wind turbine platforms, focusing on aerodynamics and hydrodynamics; however, research on the dynamics of multi-wind-turbine platforms—an extension of single floater designs—is limited. Analyzing the dynamic properties of an offshore floating multi-wind turbine platform (OFFWIND) is complex due to the intricate interaction between aerodynamics and hydrodynamics. Accurately predicting these loads is crucial for evaluating how they affect the power generation of the OFFWIND. This research aims to fill the gap regarding the dynamic analysis of OFFWIND [11] and assess the impact of blade pitch control on the performance of wind turbines.

Lefranc and Torud [2] analyzed the feasibility, design, and costs of a semi-submersible platform equipped with three wind turbines. Their study highlighted considerable power losses caused by interactions between upwind and downwind turbines at low wind speeds, whereas the effect was negligible at higher wind speeds. The economic considerations of the multi-wind turbine approach were found to be comparable to those of single platforms, especially regarding energy production costs. Hu et al. [3] validated their numerical models experimentally using a 1:50 scale model of a semi-submersible floating multi-wind turbine platform. They analyzed linear wave-body interactions through the potential flow method and measured pressure distributions on the platform caused by waves using a CFD model. Kim et al. designed a large semi-submersible platform equipped with four wind turbines located at each corner, gathering hydrodynamic force data using ANSYS AQWA, which was then applied in GL-DNV Bladed [12]. In South Korea, a square-shaped semisubmersible multi-unit floating offshore wind turbine (MUFOWT) platform is equipped with four 3 MW wind turbines positioned at each corner, complemented by 24 power absorber-type wave energy converters (WEC). Model tests were conducted on a scaled MUFOWT model by a research institute in Korea, without considering the dynamic motions of the fixed wave energy converters [1, 4, 5]. Lee et al. [6] examined multi-body hydrodynamic interactions in the frequency domain for a hybrid floating offshore wind-wave platform. Jang and colleagues [7] demonstrated that incorporating lightweight heave plates can effectively reduce pitch and heave motions in a multi-unit offshore floating wind turbine. Studies [9, 10] investigated a triangular-shaped floating platform with three wind turbines at its corners, focusing exclusively on hydrodynamic loads to assess the motion and elastic responses of the structure. Bae et al. [13] conducted a comprehensive dynamic analysis that included the blades, towers, floater, drivetrains, and mooring system of a multi-unit floating offshore wind turbine, though their study did not consider wake effects between the turbines.

This chapter details the research findings on designing a semi-submersible floating platform intended for hosting multiple wind turbines. The configuration model of the platform took into account the wake effects among the wind turbines to determine the wake expansion and velocity reductions downstream of the turbines [11]. The advanced platform, featuring five BMW wind turbines, is examined using ANSYS Fluent to study velocity profiles, pressure variations, and turbulence intensity patterns. This analysis employs the unsteady Reynolds-averaged Navier-Stokes (RANS) equations combined with the k-w shear stress transport (SST) turbulence model. Aerodynamic forces acting on the wind turbines are calculated and incorporated into the hydrodynamic framework in ANSYS Aqwa. This integration evaluates added mass, response amplitude operators (RAOs), pressure distributions, motion responses, and wave and viscous damping effects using the potential flow diffraction/radiation approach. The dynamics of the OFFWIND system are modeled by accounting for all forces acting upon it. A state-space representation of the governing equations is employed to perform simulations in MATLAB Simulink. This study implements an adaptive control algorithm, previously developed for its robustness to parameter variations [14], to manage blade pitch control. The impact of this adaptive control [14] on the blade pitch angle's performance is analyzed under wind conditions exceeding the rated speed. Key performance metrics evaluated include generated power output, platform surge, heave, and pitch motions. To verify its effectiveness, the adaptive controller is compared against a standard proportional-integral (PI) controller.

The remainder of the chapter is structured as follows and described in the flowchart below. Section 2 offers a description of the model. Section 3 presents the governing equations, numerical setup, and results related to the aerodynamics and hydrodynamics of the OFFWIND. Section 4 discusses dynamics and the impact of adaptive blade pitch control on the OFFWIND's performance. Finally, Section 5 provides conclusions and recommendations for future research (**Figure 1**).

2. Model description

The OFFWIND platform, modeled in SolidWorks with a focus on wake analysis, is a semi-submersible structure that features five 8 MW wind turbines affixed to columns that are linked by pontoons, as demonstrated in **Figure 2**. This platform is

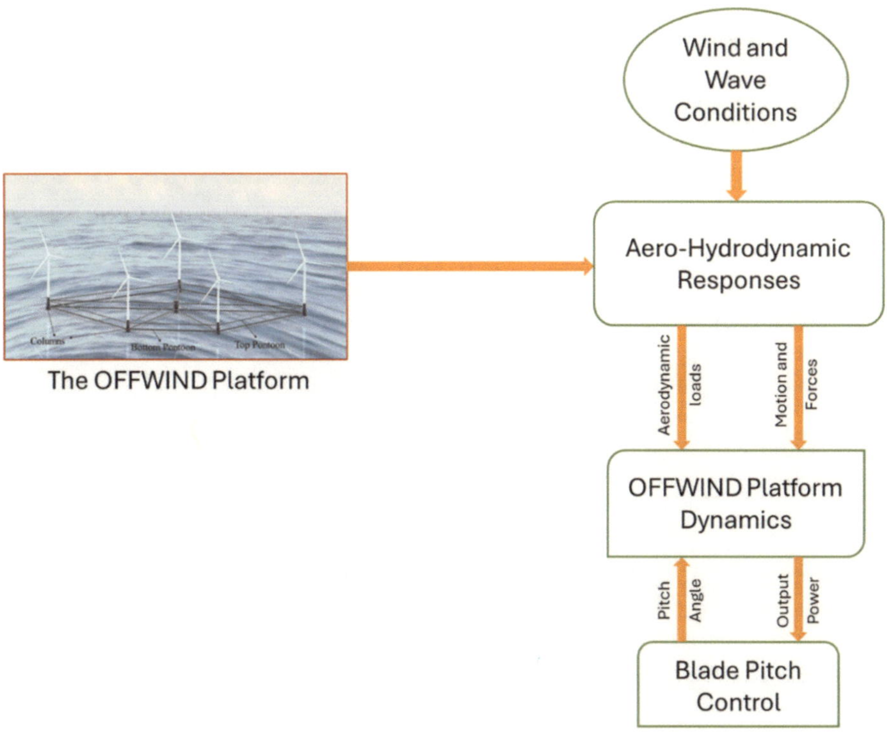

Figure 1.
The framework of the analysis and control of the OFFWIND.

engineered to reduce the wake impacts among the wind turbines and aims to produce a power output that closely matches the rated capacity. The wind turbine utilized for this study is based on the LEANWIND 8 MW reference model intended for offshore wind energy projects [15].

The design specifications for the semi-submersible platform and wind turbine are outlined in **Table 1**. To support the 8 MW wind turbine tower base with appropriate margins, the cylindrical column is designed with a height of 30 meters and a diameter of 15 meters. The diameters of the top and bottom pontoons are set at 2 meters and 3.5 meters, respectively. The platform is anchored to the seabed via a central turret system, utilizing catenary mooring lines connected to the seabed through embedded anchors.

Parameter	Value
Rating of the wind turbine	8 MW
Diameter of the rotor	164 m
Height of the hub	110 m
Nominal speed of the rotor	10.5 rpm
Wind speeds (cut-in, rated, cut-out)	4, 12.5, 25 m/s
Mass of the wind turbine	1,038,000 kg
OFFWIND mass with ballast	37,839,000 kg
OFFWIND roll inertia	8.667×10^{11} kg m^2
OFFWIND pitch inertia	4.452×10^{11} kg m^2
OFFWIND yaw inertia	1.209×10^{12} kg m^2
Number of mooring lines	4
Mooring line length	600 m

Table 1.
Properties of the OFFWIND.

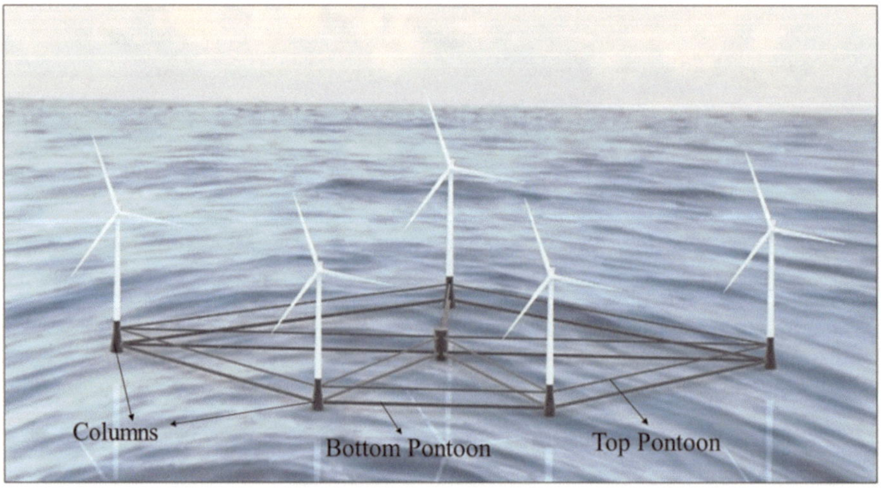

Figure 2.
Multi-wind-turbine platform.

3. Aero-hydrodynamics analysis

3.1 Aerodynamics

The continuity equation and unsteady Reynolds-averaged Navier-Stokes (RANS) equations, combined with the shear stress transport (SST) k − ω turbulence model [16], are formulated to describe the transient, incompressible and viscous fluid flow as follows.

$$\nabla.\boldsymbol{v} = 0 \tag{1}$$

This equation gives divergence of the velocity field is equal to zero.

$$\frac{\partial \overline{\boldsymbol{v}}}{\partial t} + (\overline{\boldsymbol{v}}.\nabla)\overline{\boldsymbol{v}} = -\frac{1}{\rho}\nabla\overline{p} + \nu\nabla^2\overline{\boldsymbol{v}} - \nabla.\left(\overline{\boldsymbol{v}'\,\boldsymbol{v}'}\right) \tag{2}$$

The above equation gives the change in the time-averaged velocity over time, plus the convective term of the velocity field, is equal to the negative gradient of pressure divided by fluid density, plus the kinematic viscosity times the Laplacian of the velocity field, minus the divergence of the Reynolds stress tensor.

$$\frac{\partial}{\partial t}(\rho k) + \nabla.(\rho k \boldsymbol{v}) = \nabla.[\Gamma_k \nabla k] + \tilde{P}_k - D_k \tag{3}$$

The time rate of change of the product of density and turbulent kinetic energy, augmented with the divergence of the product of density and the velocity field, is equal to the divergence of the effective diffusivity of turbulent kinetic energy times the gradient of kinetic energy, plus the turbulence production term minus the dissipation term.

$$\frac{\partial}{\partial t}(\rho\omega) + \nabla.(\rho\omega\boldsymbol{v}) = \nabla.[\Gamma_\omega \nabla\omega] + P_\omega - D_\omega + Y_\omega \tag{4}$$

The time rate of change of the product of density and specific dissipation rate, along with the divergence of the product of density and the velocity field, equals the divergence of the effective diffusivity of the specific dissipation rate multiplied by the gradient of the dissipation rate, plus the turbulence production term, minus the dissipation term, added to the cross-diffusion term.

In these equations, the variable $\boldsymbol{v}$ represents the flow field velocity, p is pressure, ν is the kinematic viscosity, and ρ is the fluid density. The effective diffusivity terms Γ_k and Γ_ω relate to both turbulent kinetic energy k and specific dissipation rate ω. $\tilde{P}_k$ and P_ω are the turbulence generation terms and D_k and D_ω are the turbulence dissipation terms. Y_ω is the cross-diffusion contribution originating from a combination of standard $k-\varepsilon$ and $k-\omega$ models [17–22]. The RANS equations can analyze the intricate dynamics within a wind turbine rotor's flow fields [23].

3.1.1 Numerical setup

This research involves solving the unstable RANS equations to simulate the rotor's rotation and its interaction with the surrounding airflow using ANSYS Fluent. The simulation employs a pressure-velocity coupling technique based on the SIMPLE (semi-implicit method for pressure-linked equations) algorithm. A predictor-corrector framework ensures the coupling between the continuity and momentum equations. For modeling convection terms in the flow equations, a second-order upwind scheme is utilized, while a second-order implicit approach is applied for temporal discretization during transient analysis.

The computational domain, measuring 6D(X) × 6D(Y) × 3D(Z), is constructed in ANSYS Design Modeler following the import of the OFFWIND model from SolidWorks. The mesh, generated using the ANSYS meshing module, consists of unstructured tetrahedral elements created through a patch-conforming algorithm chosen for its reliable convergence behavior. The mesh contains a total of 10.5 million cells, encompassing the entire OFFWIND system, as illustrated in **Figure 3**. An overset mesh methodology is applied to manage the relative motion between rotating components (rotors) and stationary structures (tower, nacelle, and platform), shown in **Figures 4** and **5**. The layout includes two upstream turbines (T1 and T2) in the first row, two additional turbines (T3 and T4) in the second row, and a downstream turbine (T5) positioned at the rear.

A fixed wind velocity of 11 m/s is specified for the inlet boundary condition, with default turbulence intensity set at 5% and a viscosity ratio of 10%. The outlet boundary condition is set to atmospheric pressure. To simulate the moving rotors of the five wind turbines placed on the platform, a multiple-moving reference frame with a sliding mesh technique is employed. The rotors are assigned the nominal rotational speed in the cell zone conditions. A no-slip boundary condition is enforced on the blade surfaces. Transient simulations are performed with a step size of 0.1 seconds, equating to a 6° blade rotation per step. The simulation spans 120 seconds of runtime, necessitating 40,000 iterations to complete.

3.1.2 Results

This section presents the pressure, velocity, and turbulence intensity contours of the OFFWIND, focusing primarily on the downstream wind turbine, which operates within the partial wake of two upstream turbines. Analyzing the downstream turbine

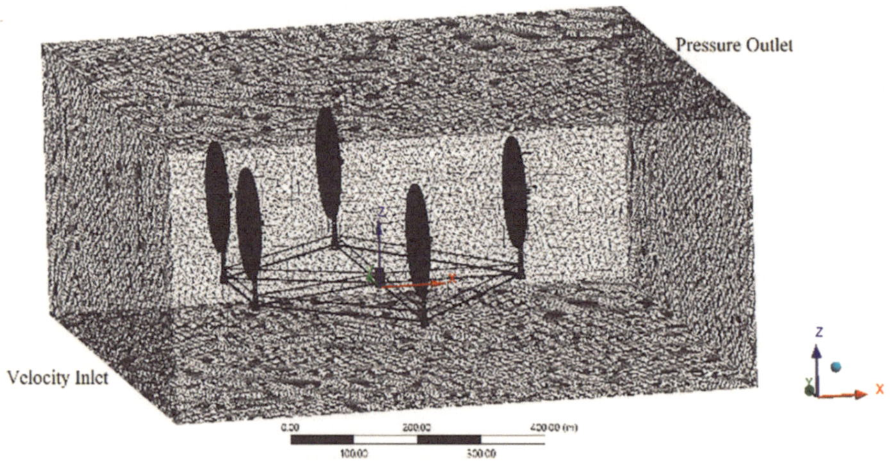

Figure 3.
Computational domain model.

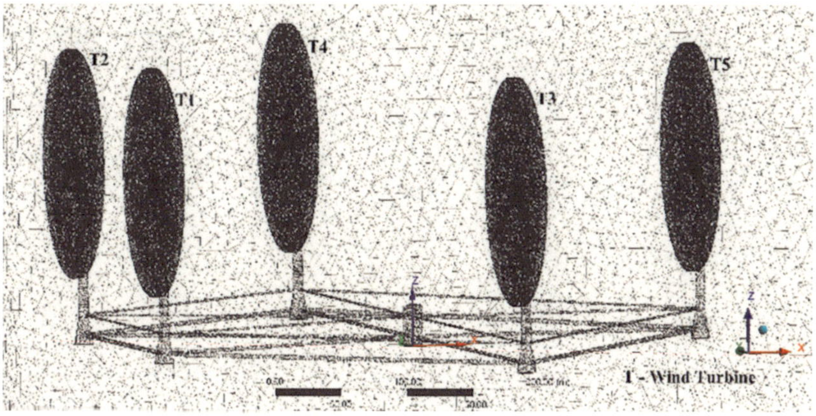

Figure 4.
Full configuration of OFFWIND with overset regions.

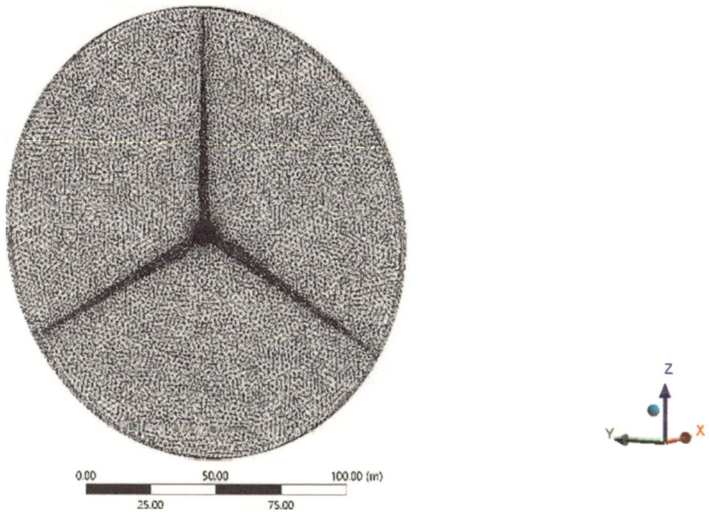

Figure 5.
The discretized representation of the rotating area.

is crucial for assessing the overall aerodynamic performance of the OFFWIND. **Figure 6** illustrates the pressure distribution along the rotor, depicting both the pressure and suction surfaces. As anticipated, a high-pressure area is visible on the pressure side, while a low-pressure area is present on the suction side. The difference in pressure generates the necessary lift, contributing to the blade's rotation. The velocity contours for the downstream turbine at various blade positions are illustrated in **Figure 7**. The interaction between the rotating blade and the tower generates fluctuations in velocity magnitudes and alters flow patterns around both structures, as illustrated in the figure.

Figure 8 depicts the turbulent intensity profiles around the wind turbines. For the upstream turbines, wake vortices lead to an increase in both the extent and magnitude of turbulence intensity in their wake. Meanwhile, the downstream turbine experiences a further escalation in turbulence intensity due to the combined effects of wake vortices generated by both the upstream and downstream turbines. This also indicates

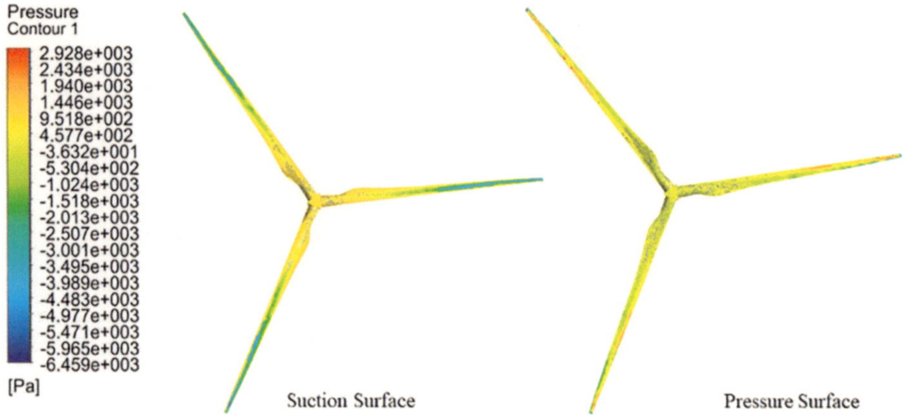

Figure 6.
Pressure curves along the rotor surface.

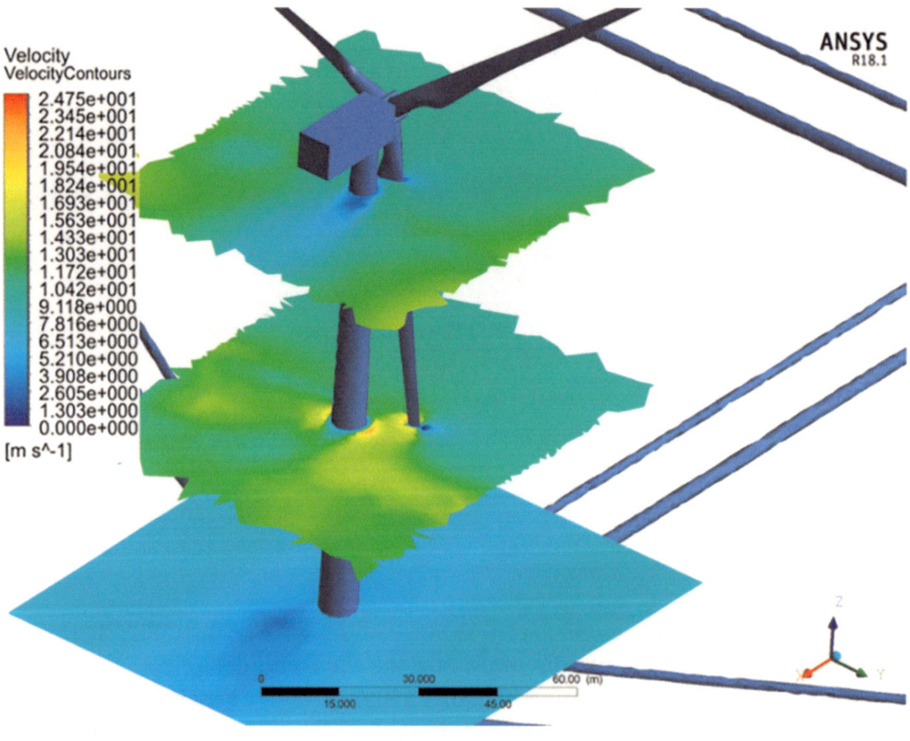

Figure 7.
Velocity contours of the downstream turbine.

that the second-row wind turbines are not affected by the wakes of the upstream turbines, thanks to their optimized arrangement based on prior research [11].

Aerodynamic thrust and torque are key parameters for evaluating the aerodynamic performance of a wind turbine. The thrust force (F) represents the total force acting perpendicular to the rotor plane, while the aerodynamic torque (T) is defined as the total moment acting along the turbine's axis of rotation [18, 24]. These parameters can be defined as follows.

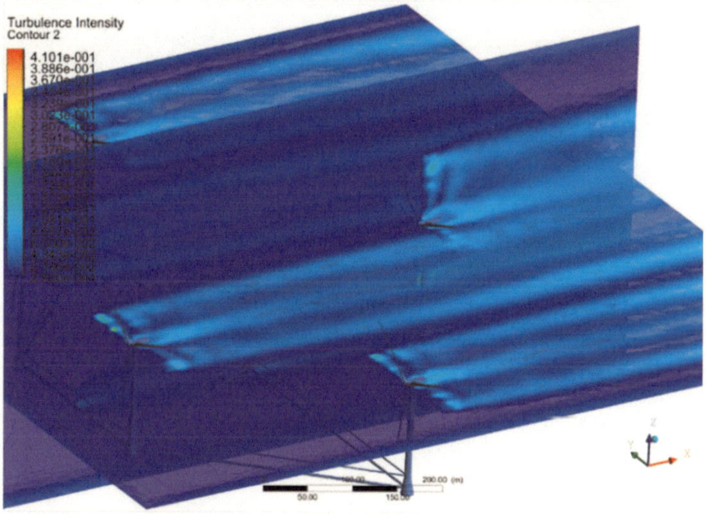

Figure 8.
Turbulence intensity contours for OFFWIND.

$$F = \oint p\boldsymbol{n}.\boldsymbol{dS} \tag{5}$$

$$T = \oint (\boldsymbol{r} \times p\boldsymbol{n}).\boldsymbol{dS} \tag{6}$$

where $\boldsymbol{n}$ represents the normal direction vector of the rotor plane, $\boldsymbol{dS}$ denotes the infinitesimal surface area vector, and $\boldsymbol{r}$ is the position vector from the center of rotation to the surface. The time history of aerodynamic thrust forces acting on the wind turbines is shown in **Figure 9**. The downstream turbine experiences greater aerodynamic forces compared to the upstream turbines due to increased turbulence caused by the partial wake effect generated by the upstream turbines, as highlighted in the magnified portion of the graph.

3.2 Hydrodynamics

When it comes to hydrodynamic analysis, the selection of the mathematical model is influenced by the structure's size, the height of the waves, and the wavelength of the incoming waves. For this hydrodynamic analysis, the first-order potential flow is utilized to address diffraction and radiation waves. The velocity potential resulting from the incident waves (ϕ_I), diffraction waves (ϕ_D), and radiation waves (ϕ_R) can be expressed [19, 25, 26] as

$$\phi = \phi_I + \phi_D + \sum_{j=1}^{6\,DOF} \zeta_j \phi_{Rj} \tag{7}$$

$$\phi_I = -\frac{Ag}{\omega} e^{\kappa z} e^{i(\kappa(x\cos\beta + y\sin\beta))} \tag{8}$$

where ζ_j refers to the displacement due to motion, A is the wave's amplitude, β indicates the direction in which the wave propagates, ω represents the wave

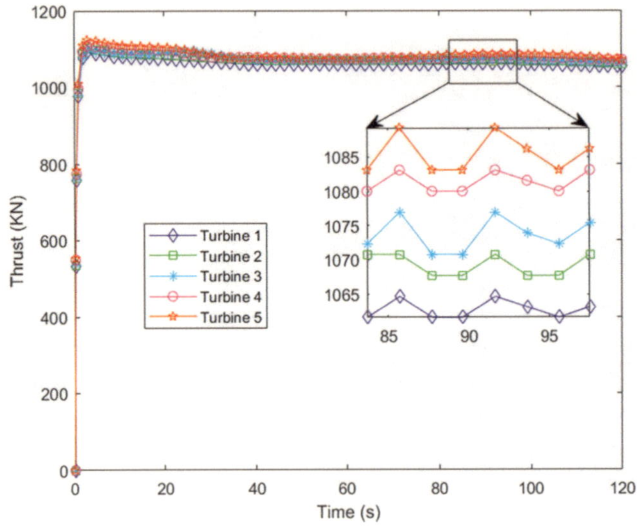

Figure 9.
Aerodynamic thrust on the rotor.

frequency, κ denotes the wavenumber, and g is the acceleration due to gravity. The structure's motion displacements are evaluated using the following motion equations.

$$\sum_{j=1}^{6} \zeta_j \left(-\omega^2 \left(A_{ij} + M_{ij}\right) - i\omega B_{ij} + C_{ij}\right) = X_i \tag{9}$$

where C_{ij} is the hydrostatic restoring force, A_{ij} and B_{ij} are the added mass and radiation damping respectively, M_{ij} is the generalized mass of the offshore floating body, and X_i is the wave excitation force.

3.2.1 Numerical setup

In this analysis, the 3D diffraction/radiation module in ANSYS Aqwa is employed within the frequency domain to calculate the first-order wave forces and hydrodynamic coefficients. These findings are subsequently utilized to determine the platform's response over time under varying wind and wave conditions. To account for the non-linearity of the system, direct numerical integration of the equations of motion is used in the time domain analysis. The hydrodynamic characteristics of the OFFWIND are presented in **Table 2**.

Using ANSYS Aqwa, mooring cables for floating platforms can be effectively simulated. These mooring lines contribute to increased damping of the platform by exerting drag forces on the structure. The reaction forces from the mooring lines are determined through cable dynamics during time response analysis. For this OFFWIND project, there are four identical catenary mooring lines that are fixed at one end to the center of the platform, while the other end is anchored to the seabed using fixed anchor points, as depicted in **Figure 10**. Each non-linear catenary mooring line is modeled using 100 elements. The lines are symmetrically positioned relative to the X-axis. The specific parameters of the mooring lines are outlined in **Table 3.**

Parameter	Value
Depth of the water	250 m
Draft	20 m
Platform's freeboard	10 m
Center of buoyancy above the keel	7.55 m
Center of gravity above the keel	24 m
Volumetric displacement	36,915.34 m^3

Table 2.
Specifications of the OFFWIND.

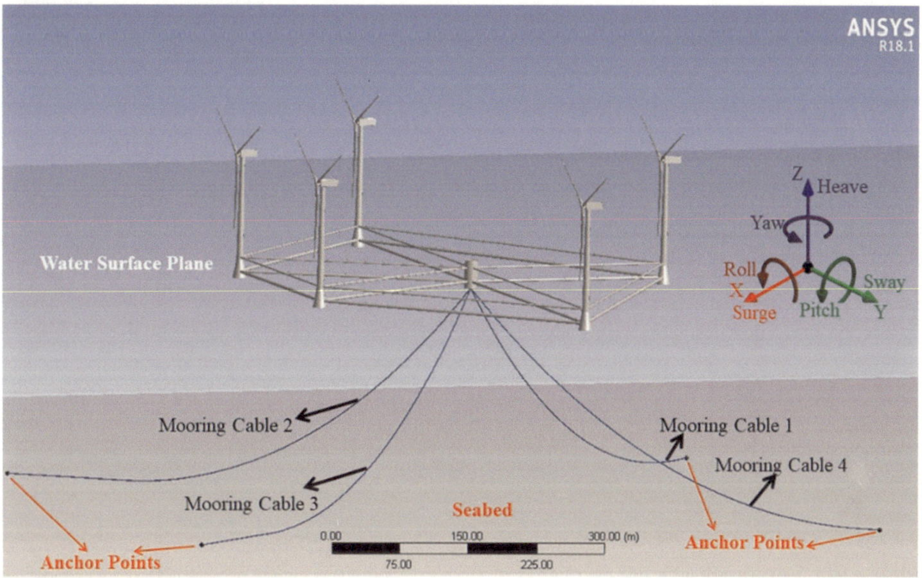

Figure 10.
OFFWIND with catenary mooring lines.

A selection of wave directions ranging from 180° to −180° with an interval of 45°, is utilized for the analysis of hydrodynamic diffraction. The spectrum of JONSWAP wave [27] depicted in **Figure 11**, is employed to simulate the asymmetrical waves under the conditions, which include a substantial wave height (H_s) of 4 m, a wave period of 9 s, and a peak parameter (γ) of 3. It is represented as.

$$S(\omega) = 5A_\gamma \left(\frac{H_s}{4}\right)^2 \frac{\omega_p^4}{\omega^5} e^{\left(-\frac{5}{4}\left(\frac{\omega_p}{\omega}\right)^4\right)} \gamma^{exp\left(-\frac{1}{2}\left(\frac{\omega-\omega_p}{\sigma\omega_p}\right)^2\right)} \tag{10}$$

where $\omega_p = \frac{2\Pi}{T_p}$ peak-spectral frequency, $A_\gamma = 1\text{–}0.287\, ln\,(\gamma)$ is a normalizing factor, σ is a spectral width parameter with $\sigma = 0.07$ for $\sigma \leq \omega_p$ and $\sigma = 0.09$ for $\sigma > \omega_p$, and T_p is the peak wave period. The incident wave's frequency range for the simulation extends from 0.1 to 2.18337 rad/s, during which the structure's hydrodynamic parameters are assessed.

Parameter	Value
Number of mooring lines	4
Length	600 m
Unit length's mass	120 kg/m
Maximum tension	7.5 MN
Stiffness, EA	900 MN
Added mass coefficient	1
Equivalent cross-section area	0.01 m^2
Longitudinal drag coefficient	0.025
Equivalent diameter	100 mm
Transverse drag coefficient	1

Table 3.
Properties of the mooring lines.

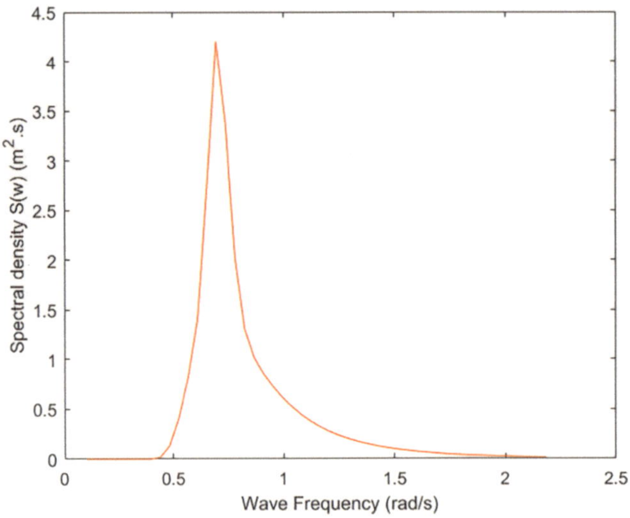

Figure 11.
JONSWAP wave spectrum.

The radiation forces in the time-domain analysis are calculated using the convolution technique. Aerodynamic loads, evaluated at a steady wind speed of 11 m/s, are imported into ANSYS Aqwa to simulate the platform's motion under the combined influence of wind and wave forces. The time-history responses are generated for a wave heading angle of 0° (aligned with the X-axis) within the hydrodynamic response module of Ansys Aqwa over a duration of 600 seconds, using a time-step size of 0.1 seconds.

3.2.2 Results

This section presents the pressure contours, frequency domain, time-domain dynamic responses, and mooring line tensions of the OFFWIND. **Figures 12–14** illustrate the frequency response amplitude operator (RAO) for the platform's surge (X), heave (Z), and pitch (RY) motions. These responses are depicted for wave direction ranges from 180° to −135° with an interval of 45° based on the hydrodynamic diffraction analysis. As illustrated in the figures, the platform responses for wave directions of −180° and 0°, −90° and 90°, −135° and − 45°, and 45° and 135° respectively are essentially the same.

The response to surges at wave-heading angles of 0° and 90° respectively is greater because of the wave forces acting in those directions. The heave response remains fairly uniform across various wave directions, with the maximum peak observed at a wave-heading angle of 90°. In contrast, the platform's pitch response is most significant at wave-heading angles of 0° and 180°.

Figure 15 presents the contour plot of the pressure distribution on the platform at a frequency of 0.99287 rad/s, with an incident wave amplitude of 2 m and a wave

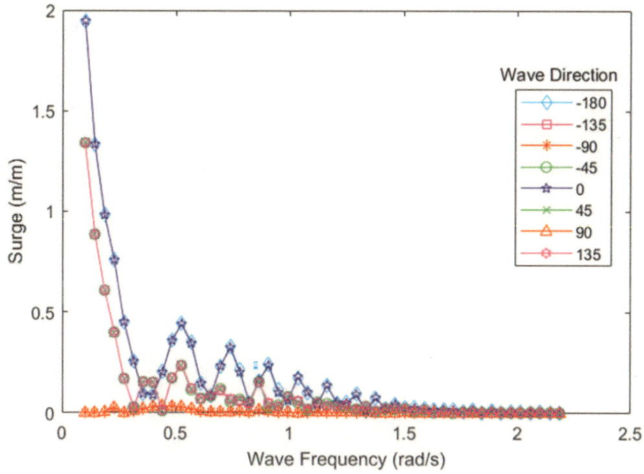

Figure 12.
The frequency response of surge motion.

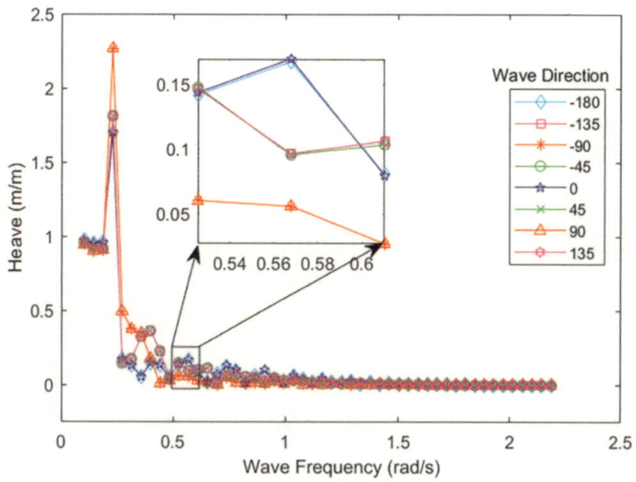

Figure 13.
The frequency response of heave motion.

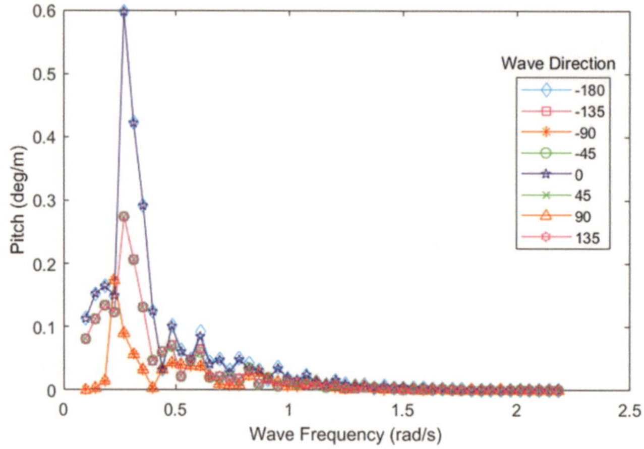

Figure 14.
The frequency response associated with pitch motion.

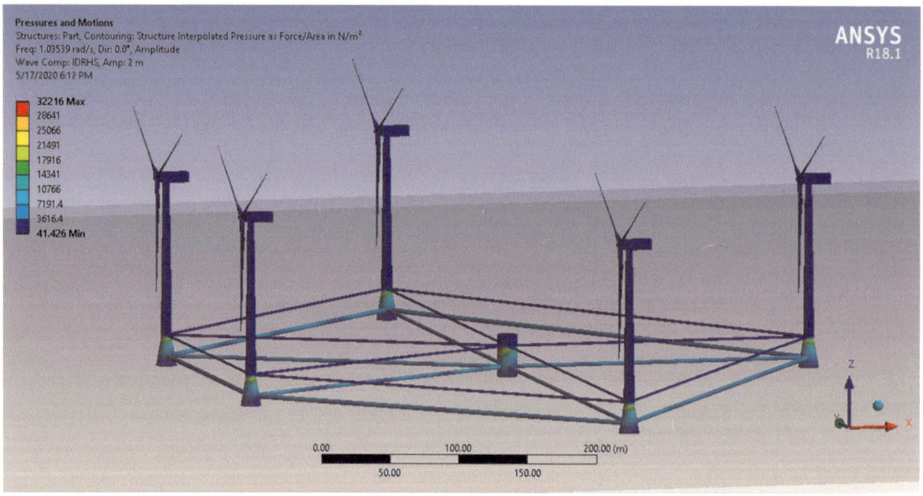

Figure 15.
Pressure contours on the OFFWIND.

direction of 0°. The highest pressure rates are observed at the surface of the water, where it interacts with the platform, as illustrated in **Figure 15**.

The response amplitude operators (RAO) for the OFFWIND under the influence of wind and wave conditions are displayed in **Figures 16–18**. The forces produced by both wind and waves significantly impact the platform's motion. The figures indicate that the six degrees of freedom (DOF) motions of the platform are periodic, albeit with varying periods for each motion.

The platform undergoes an average surge motion of 8.58 m, driven by the combined thrust force from the wind turbines and the waves traveling in the same direction. The tension in the mooring lines counteracts this motion.

The pitch and heave motions of the platform significantly influence the wind turbine's power generation, as these motions are influenced by wave energy. The average heave response is 0.267 m, while the average pitch response is 0.291 degrees,

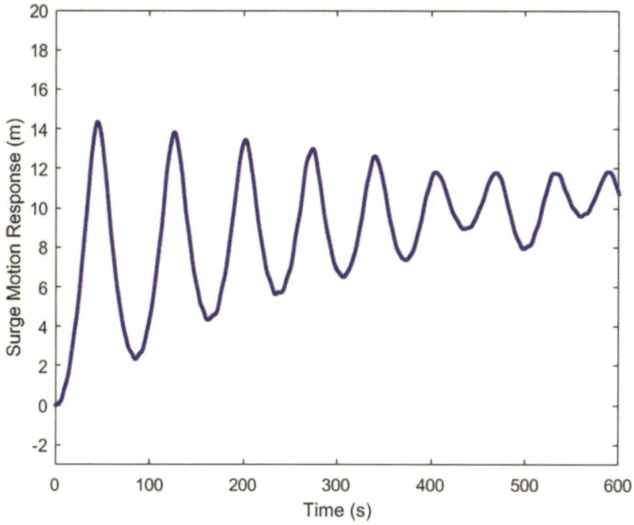

Figure 16.
Time histories of the surge motion.

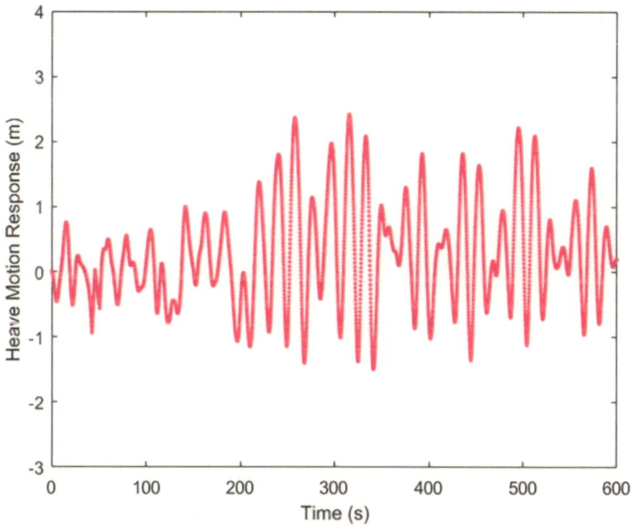

Figure 17.
Time histories of the heave motion.

which may be equal, but their maximum amplitudes are different at 2.423 m and 1.123 degrees. Utilizing heave plates can help mitigate the amplitudes of pitch and heave motions [7, 28].

Figure 18 presents the time histories of pitch motion.

Figure 19 illustrates the time histories of tensile forces in the catenary mooring lines. The tension forces in cables 1 and 4, as well as in cables 2 and 3, show similar patterns because the mooring lines are symmetrically distributed around the X-axis, while wind and wave directions align with the X-axis. As the mooring tensions are influenced by the surge motion response, it is evident from the figure that the front cables (2 and 3) experience greater force to counteract the aerodynamic thrust force.

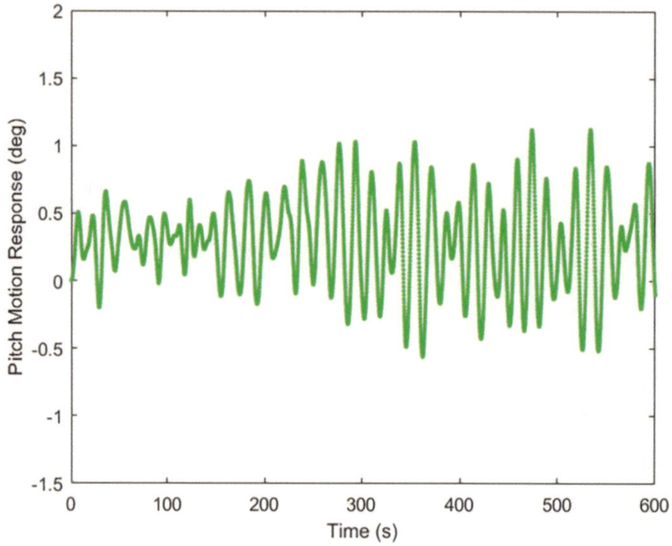

Figure 18.
Time histories of the pitch motion.

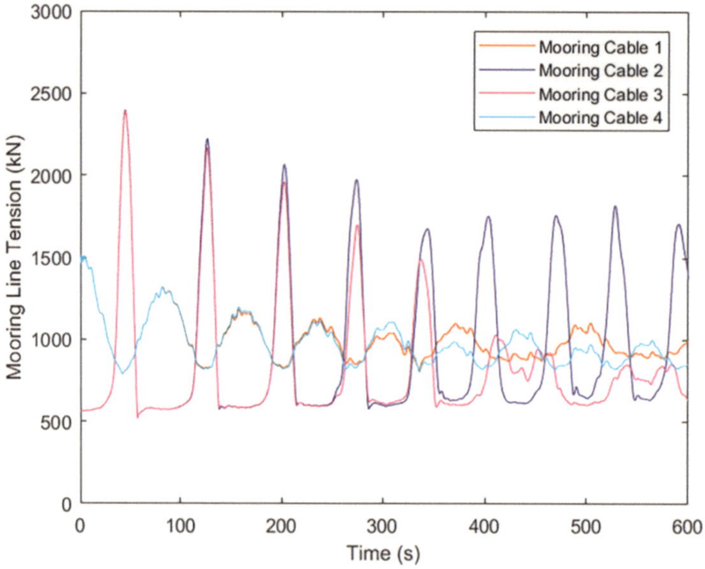

Figure 19.
Mooring line tensile forces.

4. Blade pitch control of OFFWIND

The dynamics of the system are established based on a rigid platform assumption represented by 6 DOF. Utilizing Newton's second law, the governing equation for the interconnected platform and wind turbines can be expressed as.

$$(\boldsymbol{M} + \boldsymbol{A})\ddot{\boldsymbol{u}} = \sum_{T=1}^{5} \boldsymbol{F}_{AD}^{T} + \boldsymbol{F}_{HS} + \boldsymbol{F}_{HD} + \boldsymbol{F}_{M} \tag{11}$$

where $\boldsymbol{A}$ is an added inertia matrix, $\boldsymbol{M}$ is a generalized inertia matrix containing mass and moment of inertia components, $\sum_{T=1}^{5} \boldsymbol{F}_{AD}^{T}$ is the summation of the aerodynamic forces from the five wind turbines, $\boldsymbol{F}_{HD}$ are the hydrodynamic forces, $\boldsymbol{F}_{HS}$ are the hydrostatic forces, $\boldsymbol{F}_{M}$ is the force in mooring line and $\ddot{\boldsymbol{u}}$ is the acceleration vector. The aerodynamic thrust force (T) acting on the rotor blades of the wind turbine and the aerodynamic torque (Q) that the wind exerts on the wind turbine rotor are stated as.

$$T = \frac{1}{2}\rho_{air}A\ C_T(\lambda, \beta)v^2 \tag{12}$$

$$Q = \frac{1}{2}\rho_{air}AR\ C_Q(\lambda, \beta)v^2 \tag{13}$$

where A is the rotor-swept area, ρ_{air} is the density of the air, R is the rotor radius, C_T is the aerodynamic thrust, C_Q is the torque coefficient as a function of the tip speed ratio, λ, and pitch angle, β, and v is the wind speed. The thrust force is directed horizontally along the wind's path, while the torque acts as a horizontal vector aligned with the wind, based on the assumption that the rotor remains continually oriented into the wind [29]. The hydrostatic restoring forces can be described in the following manner.

$$\boldsymbol{F}_{HS} = -\boldsymbol{K_h u}(t) \tag{14}$$

$$\boldsymbol{K_h} = \left[C_{ij}\right] \quad i,j = 1, 2, \ldots 6 \tag{15}$$

where the hydrostatic restoring coefficient matrix is denoted by $\boldsymbol{K_h}$ which contains the values of $C_{33} = -\rho_s g A_0$, $C_{44} = -mgGM_X$, $C_{55} = -mgGM_Y$ and the other elements are set to zero and $\boldsymbol{u}$ represents the displacement vector. m is the mass of the body, A_0 refers to the waterplane area, ρ_s is the seawater density, GM_X and GM_Y represent the metacentric heights in X and Y directions, respectively and g is gravity [30, 31]. The hydrodynamic force acting on the support platform is formulated as

$$\boldsymbol{F}_{HD} = \boldsymbol{F}_{HD}^{wave} - \int_0^t \boldsymbol{K}_k(t-\tau)\dot{\boldsymbol{u}}(\tau)d\tau \tag{16}$$

where the first component signifies the overall excitation load resulting from incoming waves, and the second component represents the added load input resulting from the wave radiation. The velocity vector is represented by $\dot{\boldsymbol{u}}$, $\boldsymbol{K}_k$ denotes the wave-radiation-retardation kernel matrix, and τ is a dummy variable [32]. The wave excitation force can be expressed as

$$\boldsymbol{F}_{HD}^{wave} = \sum_{n=1}^{N} A_{jn}A_n \cos\left(-\omega_n t + k_n\overline{x}\cos\theta + k_n\overline{y}\sin\theta + \Phi_{jn} + \phi_n\right) \tag{17}$$

where ω_n signifies the wave frequency, A_n represents the amplitude of the incident wave, k_n denotes the wavenumber, $(\overline{x}, \overline{y})$ is the mean horizontal offset of the structure, θ indicates the angle at which the wave is incident, ϕ_n represents the randomly generated phase, A_{jn} and Φ_{jn} denote the amplitude and phase of the nth excitation harmonic in the jth degree of freedom, respectively. A linearized model of the mooring system can describe the mooring line restoring forces in the following manner:

$$F_M = -K_m u(t) \tag{18}$$

where $\boldsymbol{u}$ indicates the displacement vector and $\boldsymbol{K_m}$ represents the mooring stiffness matrix.

4.1 State space representation

Eq. (11) is rearranged by grouping the linear terms associated with velocities, accelerations, and displacements as shown:

$$M\ddot{u} + D\dot{u} + Ku = F \tag{19}$$

where $\boldsymbol{K}$ is the stiffness matrix comprising terms linked to displacements, $\boldsymbol{D}$ is the damping matrix with terms connected to velocities, $\boldsymbol{M}$ represents the mass matrix containing all terms related to accelerations, and $\boldsymbol{F}$ is the generalized force vector that encompasses the nonlinear aerodynamic and hydrodynamic forces. Eq. (19) can be rewritten as follows:

$$\ddot{u} = -M^{-1}D\dot{u} - M^{-1}Ku + M^{-1}F \tag{20}$$

Eq. (20) can be transformed into a conventional state-space representation as given below:

$$\begin{aligned} \dot{x} &= A\,x + Bf_u \\ y &= C\,x \end{aligned} \tag{21}$$

where

$$A = \begin{bmatrix} -M^{-1}D & -M^{-1}K \\ I & 0 \end{bmatrix}; B = \begin{bmatrix} -M^{-1} \\ 0 \end{bmatrix}\ x = \begin{bmatrix} \dot{u} \\ u \end{bmatrix}; f_u = \begin{bmatrix} F \\ 0 \end{bmatrix} \tag{22}$$

In this configuration, $\boldsymbol{A}$ corresponds to the system matrix, $\boldsymbol{x}$ denotes the state vector, $\boldsymbol{y}$ is the output vector, $\boldsymbol{f_u}$ is the nonlinear input vector, $\boldsymbol{B}$ is the input matrix and $\boldsymbol{C}$ represents the output matrix.

4.2 Adaptive control

The adaptive control process created in earlier research is utilized in this simulation. The proposed control strategy was applied to 2, 3, and 5 MW wind turbines to evaluate its reliability in the face of parameter uncertainty. The findings showed that the adaptive controller is applicable to any size of the wind turbine, provided that the plant meets the almost strictly positive real (ASPR) condition [14]. The primary goal of this algorithm is to determine a control input that aligns with the output of the reference model expressed by

$$\begin{aligned} \dot{x}_m &= A_m x_m + B_m u_m \\ y_m &= C_m x_m \end{aligned} \tag{23}$$

where $\boldsymbol{A_m}$, $\boldsymbol{C_m}$, and $\boldsymbol{B_m}$ denote the state, output, and input matrices of the reference model, respectively. The adaptive control signal ($\boldsymbol{u_p}$) which consists of feedback error, model states, and model input in the feedforward path, is defined as.

$$\boldsymbol{u}_p(t) = \boldsymbol{K}_e(t)\boldsymbol{e}_y(t) + \boldsymbol{K}_x(t)\boldsymbol{x}_m(t) + \boldsymbol{K}_u(t)\boldsymbol{u}_m(t) \tag{24}$$

where $\boldsymbol{K}_e(t)$, $\boldsymbol{K}_u(t)$, $\boldsymbol{K}_x(t)$ represents the adaptive gains on error, model input, and model state respectively, and $\boldsymbol{e}_y(t)$ represents the output error. The state-space model described earlier is implemented in MATLAB Simulink to evaluate how adaptive blade pitch control influences the performance of OFFWIND under combined wind-wave conditions. **Figure 20** illustrates the adaptive control algorithm applied to the floating platform. The reference model supplies the nominal rotor speed as an input to the adaptive pitch controller, which analyzes the difference between the target and actual rotor speeds. This analysis generates the adaptive control signal (u_p), represented by the pitch angle, which is then fed into the system plant. Output parameters are assessed while maintaining a constant generator torque.

4.3 Simulation

The aim of this research is to examine the OFFWIND operating in the region where the wind speed exceeds the rated level, with the rotor speed adjusted to produce rated power through modifications to the blade pitch. Consequently, the main goal of the controller is to reduce power oscillations of the floater, while the secondary goal is to lessen the platform movements and ensure structural integrity. The controller's performance is assessed under a mean stochastic wind speed of 18 m/s, featuring a turbulence intensity of 15% and wind speed variations of 2 m/s. The wave data is simulated with irregular waves having a significant wave height of 4 m and a peak spectral period of 9 seconds for the incoming waves. The performance metrics of the OFFWIND utilizing adaptive pitch control are compared to those of the baseline PI controller to demonstrate the effectiveness of the adaptive system. The distribution of wind speed and the elevation of wave surfaces used for the 600-second runtime simulation are displayed in **Figures 21** and **22**, respectively.

This study investigates the effect of blade pitch control on power generation, platform surge, heave, and pitch motions. **Figure 23** illustrates the changes in the blade pitch angle for a specific wind turbine. The adaptive controller adjusts the pitch angle based on the discrepancy between the target rotor speed and the measured rotor speed. It is noticeable that the rate of change in pitch angle with the adaptive control is lower compared to that of the PI controller. The regulation of rotor speed for the

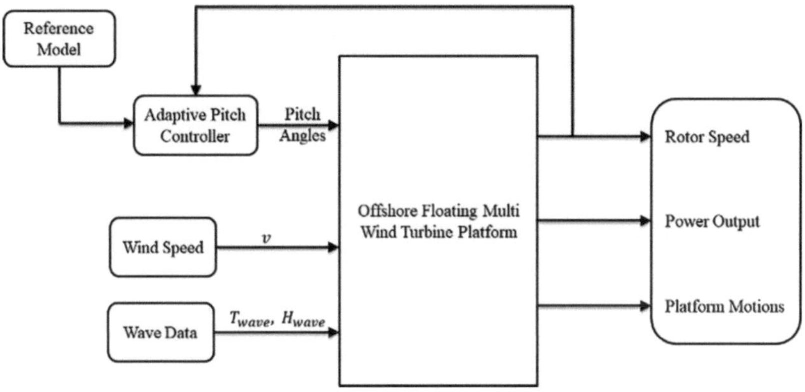

Figure 20.
Adaptive control block diagram of OFFWIND.

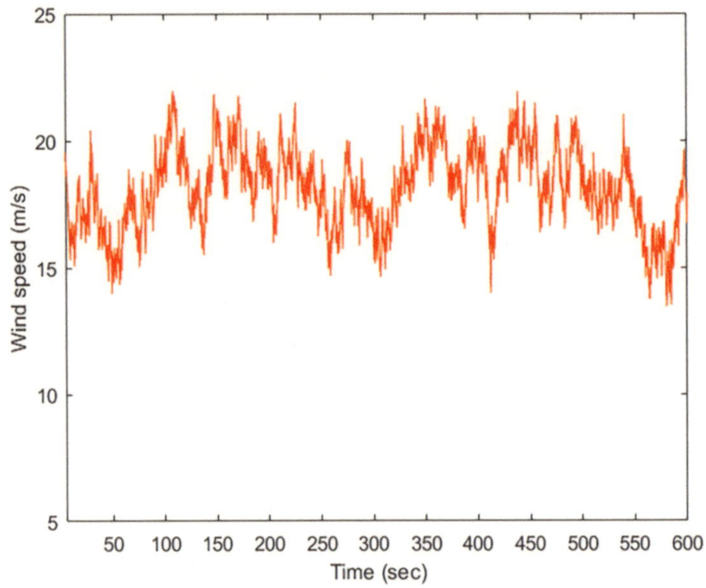

Figure 21.
Distribution of the wind speed.

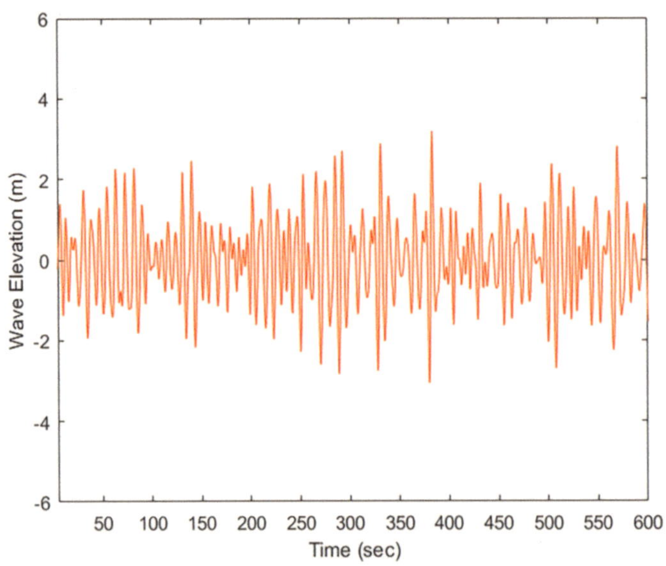

Figure 22.
Elevation of the wave surface.

chosen wind turbine in conditions above the rated wind speed is presented in **Figure 24**. The output power generated by the particular wind turbine is demonstrated in **Figure 25**. In this operating region, the power output closely follows the rotor speed due to the constant torque generated by the generator.

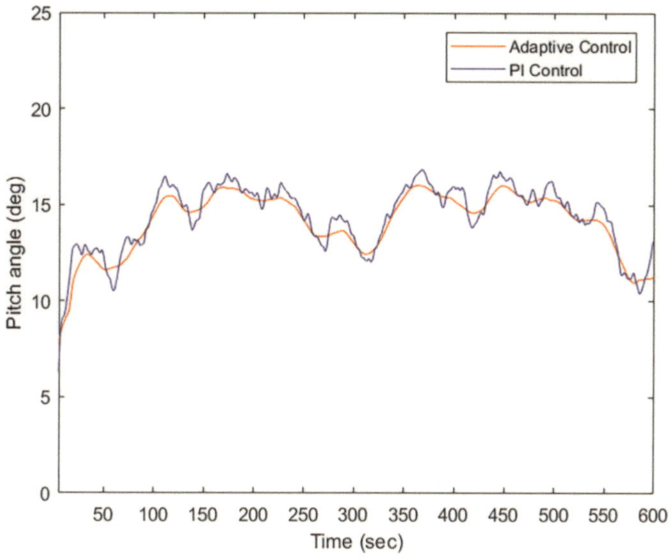

Figure 23.
Pitch angle variation of a wind turbine.

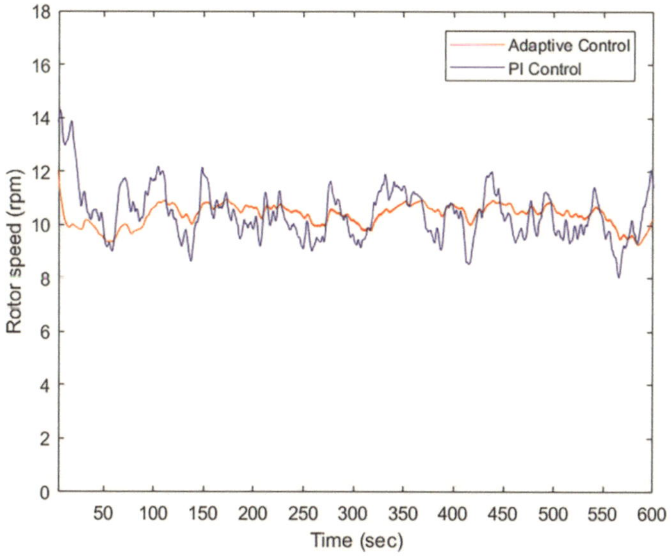

Figure 24.
Rotor speed regulation in the above-rated wind speed region.

The findings indicate that implementing adaptive control leads to a notable decrease in the variability of both rotor speed and power output compared to the PI controller. The surge, heave, and pitch motions of the floating platform are illustrated in **Figures 26–28**, respectively. The response of platform surge motion demonstrates that adaptive control effectively diminishes the peak amplitudes of surge motion in contrast to the PI controller. Both controllers exhibit a similar profile in the heave motion response. Additionally, the pitch motion of the platform shows reduced peak amplitudes with adaptive control compared to the PI controller.

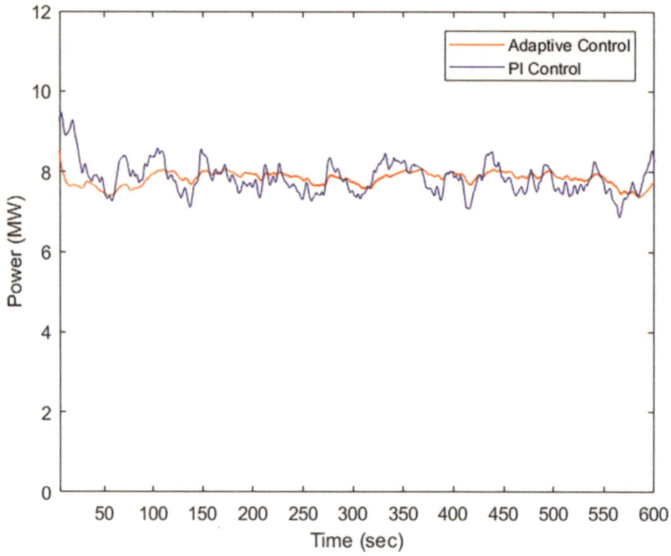

Figure 25.
The power output of a selected wind turbine.

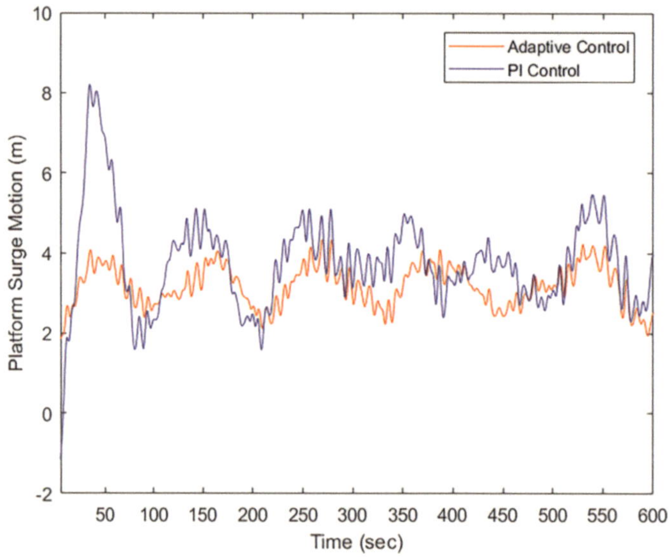

Figure 26.
Surge motion of the platform.

5. Conclusions

The study of unsteady fluid flow around the rotating rotor of a wind turbine entails a complex sequence of processes. Aerodynamic evaluations are conducted on the complete OFFWIND novel configuration, which incorporates five 8 MW wind turbines using CFD simulations to determine the aerodynamic loads on the turbines while the blades rotate simultaneously. The study discusses the contours of pressure,

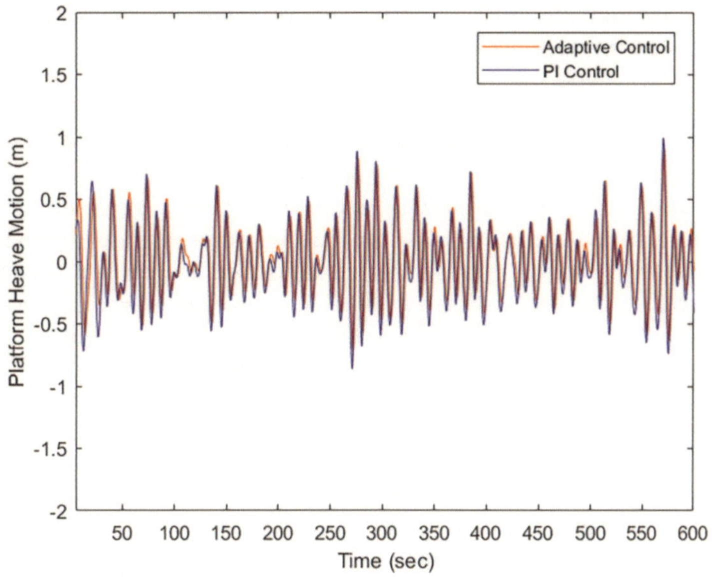

Figure 27.
Heave motion of the platform.

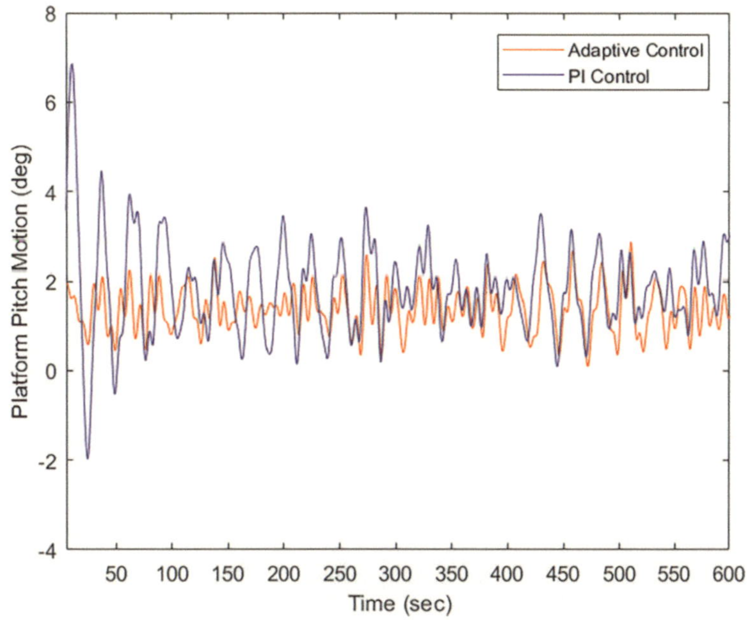

Figure 28.
Pitch motion of the platform.

velocity, and turbulence intensity for the downstream wind turbine. The time histories of aerodynamic thrust for the wind turbines indicate periodic variations in thrust force attributed to the tower shadow effect. The aerodynamic loads are subsequently integrated into the hydrodynamic module to evaluate the platform's motions in both the frequency and time domains. The structural frequency responses, analyzed for eight distinct wave heading angles, demonstrate that the responses at −180° and 0°, −90° and 90°, −135° and −45° and 45° and 135° are consistent with one another. The interpolated pressures on the structure are observed to be highest at the intersection of the water surface plane and the platform. The controller plays a crucial role in a wind turbine by managing power output and minimizing loads. The behavi or of the OFFWIND system under turbulent wind and irregular wave conditions is modeled. Blade pitch regulation is managed using an adaptive control strategy developed in prior research. The influence of this adaptive control on key performance indicators, including power output and platform dynamics, is analyzed. Results indicate that the adaptive blade pitch control outperforms the conventional proportional-integral controller in enhancing overall system performance. Additionally, the rate of blade pitch change with the adaptive control is less than observed with the PI controller. Implementing an adaptive controller results in a notable decrease in fluctuations in rotor speed and output power. The platform motions also exhibit reduced peak amplitudes in surge and pitch motion responses. In terms of heave motion response, both controllers display similar behaviors. The adaptive control can effectively be applied to the OFFWIND system when operating in winds exceeding rated speeds. This control methodology may further be analyzed for other scenarios close to the rated wind speeds. In this investigation, the simulation assumes that the wind turbines are always oriented into the wind. Future assessments could involve the development of a comprehensive yaw control system to continually align the wind turbines with the wind direction.

Author details

Srikanth Bashetty[1] and Selahattin Ozcelik[2*]

1 The University of Oklahoma, Norman, OK, USA

2 Texas A&M University—Kingsville, Kingsville, TX, USA

*Address all correspondence to: selahattin.ozcelik@tamuk.edu

References

[1] Kim HC, Kim MH, Kim KH, Hong K, Bae YH. Global performance of a square-type semi-submersible KRISO multi-unit floating wind turbine; numerical simulation vs model test. In: Presented at the the 26th International Ocean and Polar Engineering Conference; June 2016. Rhodes, Greece: ISOPE; Available from: https://www.onepetro.org/conference-paper/ISOPE-I-16-638

[2] Lefranc M, Torud A. Three wind turbines on one floating unit, feasibility, design and cost. In: Proceedings of the Offshore Technology Conference; 2–5 May 2011; Houston, TX, USA: Curran Associates, Inc

[3] Hu C, Sueyoshi M, Liu C, Liu Y. Hydrodynamic analysis of a semi-submersible-type floating wind turbine. Journal of Ocean and Wind Energy. ISOPE; 2014;**1**:7

[4] Kim KH, Hong JP, Park S, Lee K, Hong K. An experimental study on dynamic performance of large floating wave-offshore hybrid power generation platform in extreme conditions. Journal of the Korean Society for Marine Environment and Energy. 2016;**19**:7-17

[5] Kang H, Kim MH, Kim KH, Hong K. Hydroelastic analysis of multi-unit floating offshore wind turbine platform (MUFOWT). In: Proceedings of the 27th International Ocean and Polar Engineering Conference; 25–30 June 2017; San Francisco, CA, USA. Available from: https://www.onepetro.org/conference-paper/ISOPE-I-17-637

[6] Lee H, Poguluri SK, Bae YH. Performance analysis of multiple wave energy converters placed on a floating platform in the frequency domain. Energies. 2018;**11**:406

[7] Jang H-K, Park S, Kim M-H, Kim K-H, Hong K. Effects of heave plates on the global performance of a multi-unit floating offshore wind turbine. Renewable Energy. 2019;**134**: 526-537

[8] Li S, Lamei A, Hayatdavoodi M, Wong C. Concept design and analysis of wind-tracing floating offshore wind turbines. In: Proceedings of the ASME 2019 2nd International Offshore Wind Technical Conference; 3–6 November 2019; St. Julian's, Malta. NY, USA: ASME Publishing

[9] Lamei A, Hayatdavoodi M, Wong C, Tang B. On motion and hydroelastic analysis of a floating offshore wind turbine. In: Proceedings of the ASME 2019 38th International Conference on Ocean, Offshore and Arctic Engineering; 9–14 June 2019; Scotland, UK. NY, USA: ASME Publishing

[10] Lamei A, Hayatdavoodi M. On motion analysis and elastic response of floating offshore wind turbines. Journal of Ocean Engineering and Marine Energy. 2020;**6**:71-90

[11] Bashetty S, Ozcelik S. Design and stability analysis of an offshore floating multi-wind turbine platform. Inventions. 2022;**7**(3):58. DOI: 10.3390/inventions7030058

[12] Kim K, Kim H, Lee J, Kim S, Paek I. Design and performance analysis of control algorithm for a floating wind turbine on a large semi-submersible platform. Journal of Physics Conference Series. 2016;**753**:092017

[13] Bae YH, Kim M-H. The dynamic coupling effects of a MUFOWT (multiple unit floating offshore wind turbine) with partially broken blade.

Journal of Ocean and Wind Energy. 2015;**2**:2

[14] Bashetty S, Guillamon JI, Mutnuri SS, Ozcelik S. Design of a robust adaptive controller for the pitch and torque control of wind turbines. Energies. 2020;**13**(5):1195. DOI: 10.3390/en13051195

[15] Desmond C, Murphy J, Blonk L, Haans W. Description of an 8 MW reference wind turbine. Journal of Physics: Conference Series. 2016;**753**: 092013. DOI: 10.1088/1742-6596/753/9/092013

[16] Menter FR. Review of the shear-stress transport turbulence model experience from an industrial perspective. International Journal of Computational Fluid Dynamics. 2009; **23**(4):305-316. DOI: 10.1080/10618560902773387

[17] Tran TT, Kim D-H. Fully coupled aero-hydrodynamic analysis of a semi-submersible FOWT using a dynamic fluid body interaction approach. Renewable Energy. 2016;**92**:244-261. DOI: 10.1016/j.renene.2016.02.021

[18] Liu Y, Xiao Q, Incecik A, Peyrard C, Wan D. Establishing a fully coupled CFD analysis tool for floating offshore wind turbines. Renewable Energy. 2017;**112**: 280-301. DOI: 10.1016/j.renene.2017.04.052

[19] Bashetty S, Ozcelik S. Review on dynamics of offshore floating wind turbine platforms. Energies. 2021; **14**(19):6026. DOI: 10.3390/en14196026

[20] Bashetty S, Ozcelik S. Aero-hydrodynamic analysis of an offshore floating multi-wind-turbine platform—Part I. In: 2020 IEEE 3rd International Conference on Renewable Energy and Power Engineering (REPE); Edmonton, AB, Canada. NJ, USA: IEEE Publishing; 2020. pp. 1-6. DOI: 10.1109/REPE50851.2020.9253880

[21] Bashetty S, Ozcelik S. Aero-hydrodynamic analysis of an offshore floating multi-wind-turbine platform – Part II. In: 2020 IEEE 3rd International Conference on Renewable Energy And Power Engineering (REPE); Edmonton, AB, Canada. NJ, USA: IEEE Publishing; 2020. pp. 7-11. DOI: 10.1109/REPE50851.2020.9253964

[22] Bashetty S, Ozcelik S. Effect of pitch control on the performance of an offshore floating multi-wind-turbine platform. Journal of Physics Conference Series. 2021;**1828**:012055

[23] Wu C-HK, Nguyen V-T. Aerodynamic simulations of offshore floating wind turbine in platform-induced pitching motion: Numerical simulations of FOWT in coupled motions. Wind Energy. 2017; **20**(5):835-858. DOI: 10.1002/we.2066

[24] Zhang Y, Kim B. A fully coupled computational fluid dynamics method for analysis of semi-submersible floating offshore wind turbines under wind-wave excitation conditions based on OC5 data. Applied Sciences. 2018;**8**(11):Art. no. 11. DOI: 10.3390/app8112314

[25] Newman JN. Marine Hydrodynamics, 40th Anniversary Edition. Cambridge, MA, USA: The MIT Press. Available from: https://mitpress.mit.edu/books/marine-hydrodynamics-40th-anniversary-edition; [Accessed: May 03, 2020]

[26] Lin Y-H, Kao S-H, Yang C-H. Investigation of hydrodynamic forces for floating offshore wind turbines on spar buoys and tension leg platforms with the mooring systems in waves. Applied

Sciences. 2019: Art. no. 3;**9**(3). DOI: 10.3390/app9030608

[27] Hasselmann K, Olbers D. Measurements of wind-wave growth and swell decay during the Joint North Sea Wave Project (JONSWAP). Ergaenzungsheft zur Deutschen Hydrographischen Zeitschrift, Reihe A. 1973;**8**(12):1-95

[28] Yang W, Tian W, Hvalbye O, Peng Z, Wei K, Tian X. Experimental research for stabilizing offshore floating wind turbines. Energies. 2019;**12**(10): Art. no. 10. DOI: 10.3390/en12101947

[29] Antonutti R, Peyrard C, Johanning L, Incecik A, Ingram D. The effects of wind-induced inclination on the dynamics of semi-submersible floating wind turbines in the time domain. Renewable Energy. 2016;**88**: 83-94. DOI: 10.1016/j.renene.2015.11.020

[30] Ishihara T, Phuc PV, Sukegawa H. A numerical study on the dynamic response of a floating offshore wind turbine system due to resonance and nonlinear wave. In: Presented at the European Offshore Wind 2007; December 2007; Berlin. Berlin, Germany: WindFair Publishing

[31] Zambrano T, MacCready T, Kiceniuk T Jr, Roddier DG, Cermelli CA. Dynamic modeling of deepwater offshore wind turbine structures in Gulf of Mexico storm conditions. In: OMAE2006, Volume 1: Offshore Technology; Offshore Wind Energy; Ocean Research Technology; LNG Specialty Symposium. NY, USA: Hamburg, Germany, ASME Publishing; 2006. pp. 629-634. DOI: 10.1115/OMAE2006-92029

[32] Jonkman JM. Dynamics of offshore floating wind turbines—Model development and verification. Wind Energy. 2009;**12**(5):459-492. DOI: 10.1002/we.347

Chapter 3

Design Optimisation of Offshore Wind Turbine Support Structures

Lin Wang and Athanasios Kolios

Abstract

The support structure is a critical component of the wind turbine system, significantly influencing the overall performance and efficiency of the turbine, especially in offshore environments. This chapter focuses on the design optimisation of offshore wind turbine (OWT) support structures. It covers structural modelling, which is foundational in predicting how structures behave under various load conditions. It also includes discussions on design load cases and limit states, which are crucial for ensuring the structure's safety and functionality in harsh marine environments. Additionally, it explores optimisation algorithms, for example, the genetic algorithm (GA), which can effectively search for optimal solutions. A case study on the design optimisation of a typical OWT support structure is also presented, illustrating the application of these principles.

Keywords: wind turbine, support structure, design load case, design optimisation, optimisation algorithm

1. Introduction

Since the 1980s, wind energy has undergone rapid development, establishing itself as one of the world's most significant renewable energy sources. With installations increasing by over 1500% in the past 15 years, global capacity reached approximately 940 GW by the end of 2022 [1]. This expansion underscores wind energy's pivotal role in addressing growing energy demands and achieving environmental objectives. Offshore wind turbines (OWTs), in particular, offer distinct advantages over land-based turbines, including access to stronger and more consistent wind resources at sea. These benefits, coupled with technological advancements in turbine design, have driven considerable investment in offshore installations.

Support structures are essential to the safe and effective operation of OWTs, particularly given the challenging conditions of marine environments. The design of these structures must consider factors, such as water depth, seabed type, and financial constraints, as they influence stability, safety, and cost-effectiveness. Monopiles, currently the most widely used support structure across Europe, provide a robust and economical solution for shallow waters up to 30 metres deep. However, for deeper waters, alternative designs, such as jackets and floating foundations, are necessary to ensure structural stability and optimised performance.

The economic viability of offshore wind energy depends significantly on the design of support structures. Thus, optimising support structures is crucial for enhancing the competitiveness of offshore wind energy.

Developing optimised support structures for OWTs requires sophisticated modelling approaches capable of accurately predicting structural performance in complex marine environments. While traditional one-dimensional (1D) beam models offer quick and computationally efficient analyses, they often lack the precision necessary for detailed stress and local response predictions. Consequently, three-dimensional (3D) finite element analysis (FEA) models are increasingly employed in structural analysis of wind turbine structures [2, 3], providing a more comprehensive view of stress distributions and structural responses, which are essential for safe and reliable designs.

For monopile foundations, the modelling of soil-structure interaction is crucial. Accurate representation of soil properties and their effects on the structure's natural frequencies and responses under varying loads is essential for reliable design, as soil characteristics can greatly influence monopile's stability. Advanced FEA models incorporating realistic soil conditions help ensure reliable predictions for structural behaviour.

To manage the complexities inherent in offshore structural design, metaheuristic optimisation algorithms, such as genetic algorithms [4], are commonly used. These algorithms, inspired by natural selection, efficiently explore extensive design spaces without the need for gradient calculations. This approach is particularly valuable for complex engineering problems in offshore environments, enabling optimised solutions under multiple load scenarios.

This chapter explores key principles for the design and optimisation of OWT support structures, including structural modelling, load case analysis, limit states, and optimisation strategies. The chapter is structured as follows. Section 2 introduces various types of OWT support structures. Section 3 presents sources of loading. Sections 4 and 5 present design load cases and limit states, respectively. Section 6 presents structural modelling. Section 7 presents optimisation algorithms. Section 8 presents a case study on integrated design optimisation of OWT support structures.

2. Types of OWT support structures

The selection of appropriate support structures is crucial for OWT installations, as the foundations must withstand harsh marine environments, including dynamic wind, wave, and current loads. Several types of support structures have been developed to accommodate different water depths, seabed conditions, and economic considerations. The main types of OWT support structures are monopiles, jackets, twisted jackets, semi-submersible floating platforms, tension-leg platforms (TLPs), and spar platforms, as illustrated in **Figure 1**.

Each type of OWT support structure in **Figure 1** is discussed.

2.1 Monopile foundations

Monopiles are the most commonly used support structures for OWTs, especially in shallow waters. Monopiles consist of a single large-diameter steel tube driven into the seabed. They are cost-effective and relatively easy to manufacture and install, making them suitable for water depths up to 30 metres.

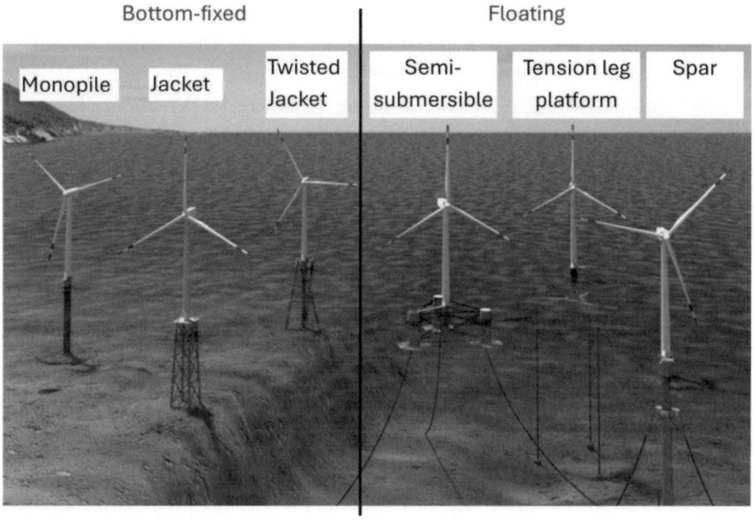

Figure 1.
Types of OWT support structures.

Monopiles are favoured for their simplicity, robust design, and efficient load transfer capabilities. However, as water depth increases, their economic viability diminishes due to the need for thicker walls and longer piles, which increase material costs. In deeper waters, monopiles can also become more prone to excessive deflection and dynamic instability.

2.2 Jacket foundations

Jacket structures are steel lattice frameworks typically used for OWTs in water depths ranging from 30 to 50 metres [5]. Originally designed for offshore oil and gas platforms, jacket foundations provide excellent structural stability by distributing loads across multiple legs anchored to the seabed with piles.

Jackets are advantageous for intermediate-depth waters, where monopiles become inefficient. Their multi-legged design reduces the dynamic response to waves and wind, making them more stable under varying conditions. Despite being more expensive and complex to manufacture and install compared to monopiles, jacket foundations offer a viable solution for intermediate-depth locations where structural stability and reduced motion are critical.

2.3 Twisted jacket foundations

Twisted jacket foundations are an adaptation of traditional jacket structures, with a twisted or angled lattice framework designed to reduce drag. The twisted geometry also helps to optimise load distribution across the structure, enhancing stability and reducing fatigue loads on critical points. This configuration is particularly advantageous in areas with high wind and wave energy, where standard jackets may experience increased drag. Twisted jackets provide an efficient solution for intermediate-depth waters, balancing cost and performance.

2.4 Semi-submersible floating platform

Semi-submersible floating platforms are used in deep waters (typically greater than 50 metres) [6], where fixed-bottom structures are impractical. These platforms consist of multiple buoyant columns connected by a horizontal platform that supports the turbine tower. Anchored to the seabed by mooring lines, semi-submersible platforms remain stable by leveraging their buoyancy and weight distribution. This type of platform is versatile, suitable for various seabed conditions, and can be towed to the installation site, reducing construction costs.

2.5 Tension-leg platform

Tension-leg platforms (TLPs) are floating structures anchored to the seabed with taut mooring lines that restrict vertical movement, allowing for a more stable platform that minimises vertical displacement caused by waves. This stability is achieved by balancing the tension in the mooring lines with the buoyancy of the platform. TLPs are particularly suited to deep waters and high-energy environments where stability is essential for turbine operation. Although TLPs have a higher installation cost due to complex mooring requirements, they provide a reliable foundation in challenging offshore settings.

2.6 Spar floating platform

Spar floating platforms are tall, slender structures that extend deep into the water, anchored by a single long cylindrical buoy that keeps the turbine upright and stable. Spar platforms are typically used in very deep waters, often exceeding 100 metres, where the long keel provides stability and reduces the impact of wave forces. Anchored with mooring lines, spar platforms are highly resistant to vertical and horizontal displacement, making them suitable for extreme offshore environments. Due to their depth requirements, spar platforms are more challenging and costly to install but offer exceptional stability and durability.

2.7 Selection criteria for support structures

The choice of support structure for an OWT is influenced by several factors, including:

- *Water depth*: Monopiles are typically used in shallow waters (up to 30 metres), while jacket foundations are more suitable for intermediate depths (30–50 metres). Floating platforms are used in waters exceeding 50 metres.

- *Seabed conditions*: The seabed composition influences the type of foundation used. Monopiles are suited to sandy or clay soils, while jacket foundations can adapt to a wider range of seabed conditions.

- *Environmental loads*: Wind, wave, and current loads dictate the structural stability requirements. In areas with high dynamic loading, jacket and floating platforms provide better stability due to their distributed load-bearing capacity.

- *Installation costs*: Monopile structures are relatively simple and cost-effective to install. In contrast, jacket and floating foundations involve more complex installations, which increase costs.

- Maintenance and decommissioning: Monopiles are relatively straightforward to maintain due to their simple design, but decommissioning can be challenging and costly, as they are deeply embedded in the seabed and require specialised equipment for extraction. Jacket structures, with their complex lattice design, are more difficult and costly to maintain and decommission, as each leg and pile must be detached from the seabed, especially in deeper waters. Floating platforms, however, offer greater flexibility for both maintenance and decommissioning; they can be towed to shore for extensive repairs and, once disconnected from mooring lines and anchors, can be removed with minimal seabed disturbance, making them a favourable option for deeper waters where decommissioning costs are higher.

3. Sources of loading

OWT support structures are subject to various environmental loads. Design standards, such as IEC 61400-3 [7] and DNVGL-ST-0126 [8], provide guidelines for key loads that must be considered when designing OWT support structures. As shown in **Figure 2**, the loads acting on the support structure (with a monopile foundation used here as an example) can be broadly classified into six categories: (1) inertial loads, (2) aerodynamic loads transferred from the rotor, (3) wind loads on the tower, (4) wave loads, (5) current loads, and (6) hydrostatic loads.

The loads shown in **Figure 2** are described in detail.

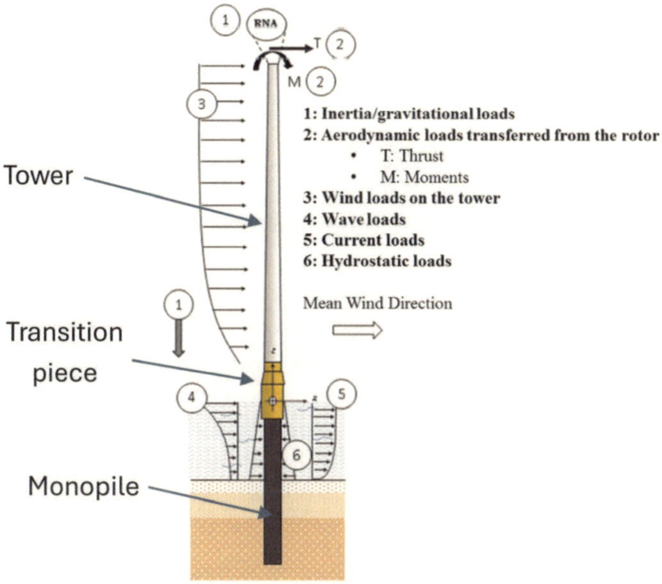

Figure 2.
Loads on OWT support structures.

3.1 Inertial loads

Inertia loads, which arise from the mass of the support structure and the RNA (rotor-nacelle assembly) mass at the top of the tower, can have a significant impact on buckling and influence the modal frequencies of the OWT support structure. Therefore, they should be considered in the structural analysis of the support structure.

3.2 Aerodynamic loads transferred from the rotor

Aerodynamic loads acting on the rotor can be transmitted to the top of the tower. For instance, the rotor thrust force T can be computed by:

$$T = \frac{1}{2}\rho V^2 C_T \pi R^2 \tag{1}$$

where ρ is the air density, V is the wind speed, C_T is the thrust coefficient, and R is the rotor radius.

3.3 Wind loads on the tower

Wind loads on the tower are generated by drag forces and depend on the wind velocity. The wind shear is typically modelled using a power law profile, expressed by the following equation:

$$v(z) = v_{ref}\left(\frac{z}{z_{ref}}\right)^{\alpha} \tag{2}$$

where v_{ref} is the reference wind speed measured at the nacelle height z_{ref}, and α is the roughness coefficient, with a typical value of 0.115 for offshore sites [9].

The wind loads along the tower are then calculated using:

$$F_{tower}(z) = \frac{1}{2}\rho C_{d,T} D(z) v(z)^2 \tag{3}$$

where $C_{d,T}$ is the drag coefficient of the tower, with a typical value of 1.0, and $D(z)$ is the external diameter of the tapered tower at height z.

3.4 Wave loads

A widely used method for calculating wave loads in shallow water is the linear (Airy) wave theory, where the surface elevation is represented by a sinusoidal shape. By applying potential theory and boundary conditions, the velocity potential can be determined, allowing the horizontal velocity and acceleration of the water particles to be derived as follows:

$$u(z,t) = \omega a \frac{\cosh k(h+z)}{\sinh kh}\sin(\omega t - kx) \tag{4}$$

$$\dot{u}(z,t) = \omega^2 a \frac{\cosh k(h+z)}{\sinh kh}\cos(\omega t - kx) \tag{5}$$

where ω is the angular frequency of the wave, a is the wave amplitude, h is the local water depth, and k is the wave number.

For slender structures like monopiles, wave loads are typically estimated using Morison's equation, which combines inertia and drag forces:

$$F_{wave}(z) = \frac{1}{4}\rho_{water}\pi D^2 C_M \dot{u}(z,t) + \frac{1}{2}\rho_{water} D C_D u(z,t)|u(z,t)| \tag{6}$$

where ρ_{water} is the water density, with a typical value of 1025 kg/m^3, C_M and C_D are the inertia and drag coefficients for the monopile, with typical values of 1.6 and 1.0, respectively [10].

3.5 Current loads

Sea current flow can generate drag forces on the support structure. The sea current velocity $u_c(z)$ from the mean sea level (MSL) to the seabed is described using an exponential profile:

$$u_c(z) = u_{c,MSL}\left(\frac{d+z}{d}\right)^{1/7} \tag{7}$$

where $u_{c,MSL}$ is the sea current velocity at MSL and d is the water depth from MSL to seabed. Assuming the current and wave are aligned, current velocity is added to the wave particle velocity in the drag component of Morison's equation in Eq. (6).

3.6 Hydrostatic loads

When the monopile is submerged in water, it experiences hydrostatic pressure on its surface. This pressure represents a constant normal load that increases linearly with water depth. The hydrostatic force F_h can be calculated as:

$$F_h = \rho_{water} g h \tag{8}$$

where g is the gravitational constant and h is the water depth.

4. Design load cases

Design load cases (DLCs) are essential for ensuring the safe and reliable operation of OWTs under various environmental and operational conditions. IEC 61400-3 [7] outlines 32 DLCs that cover a range of operational, shut-down, and fault scenarios. These load cases consider the combined effects of wind, waves, currents, and other environmental factors on the support structure. DLCs can be categorised into ultimate load cases (ULCs) and fatigue load cases (FLCs), which address extreme and cyclic loads, respectively.

4.1 Ultimate load cases (ULCs)

Ultimate load cases deal with extreme environmental conditions, such as high wind speeds and storm waves, which can occur infrequently but have the potential to cause significant damage to the turbine structure. ULCs are crucial for designing the structural integrity of wind turbine support systems to ensure they can withstand rare events.

- *Extreme turbulence model* (*ETM*): One of the key ULCs, the extreme turbulence model considers turbulence during normal turbine operation. Extreme turbulence can lead to significant aerodynamic loading on the rotor blades, resulting in high stresses on the tower and foundation. The IEC 61400-3 standard requires assessing these load cases under various wind speeds, particularly near the rated wind speed, where turbines operate at maximum power output.

- *Extreme wind shear* (*EWS*): Wind shear occurs when there are significant differences in wind speeds between the lower and upper sections of the rotor. This can lead to uneven loading on the rotor blades and the tower, increasing the risk of structural damage. Load cases for EWS consider scenarios where wind shear is at its maximum, which is especially critical for large rotor diameters.

- *Extreme wave conditions* (*EWCs*): In offshore environments, the impact of waves on the support structure must be accounted for. Extreme wave models are used to simulate the effect of waves on the foundation. These models also consider extreme water level conditions, such as storm surges, which can significantly increase the loads on the structure. For example, the 50-year return period event is a standard extreme condition considered for the ultimate design of wind turbines.

4.2 Fatigue load cases (FLCs)

Fatigue load cases address the long-term effects of cyclic loading, which can degrade the structural integrity of wind turbine components over time. OWT support structures are subjected to continuous, cyclic loads from wind and waves, making fatigue analysis critical to ensuring a long service life.

- *Normal turbine operation* (*NTO*): This is the most common operating state of the wind turbine. During NTO, the rotor spins under varying wind conditions, and the support structure experiences cyclic loading from the combination of rotor thrust and environmental forces. Fatigue damage accumulates gradually in the tower and foundation, and it is vital to assess these loads over the turbine's 20- to 25-year operational life.

- *Shut-down and start-up cycles*: Turbine shut-downs and start-ups, which occur during maintenance, emergency stops, or grid disconnections, can induce high loads on the support structure. These transient load cases must be assessed for their contribution to fatigue. The combination of wind and wave loads during shut-down, particularly in extreme weather conditions, can amplify fatigue effects, especially in the tower base and the monopile.

- *Fault conditions*: Faults, such as electrical failures or mechanical malfunctions, can result in abrupt changes in rotor speed, leading to transient loads. The IEC 61400-3 standard prescribes load cases for faults to ensure that the turbine and its support structure can withstand the resulting forces without failure. The fatigue effects of such transient loads, although brief, are important for ensuring the structure's long-term reliability.

4.3 Environmental load combinations

In OWTs, environmental loads rarely act independently. Load combinations, where wind, wave, and current forces interact simultaneously, are more realistic and must be considered during the design phase.

- *Wind-wave interaction*: The dynamic response of the turbine tower and foundation to wind and waves is a critical factor in the design process. These interactions can lead to resonant conditions, where the natural frequency of the structure aligns with the frequency of wave or wind loads, amplifying vibrations. It is essential to ensure that the turbine's operational frequencies are sufficiently separated from these resonant conditions.

- *Current loads*: In addition to wave and wind forces, sea currents can exert significant drag forces on the support structure, particularly in deep water. The IEC 61400-3 standard requires accounting for currents in combination with wave forces, as this can increase the total lateral load on the foundation. The effect of sea currents is particularly relevant for floating platforms, where the interaction between mooring lines and currents can lead to complex load scenarios.

- I*ce loads (for cold regions)*: In certain regions, ice buildup on the structure or ice floes impacting the foundation can generate additional loads. These loads must be factored into the DLCs, especially for turbines located in cold regions. Ice loading scenarios can include both static and dynamic impacts, and they require specialised models to predict the forces acting on the turbine.

4.4 Special considerations

- *Parked and idling states*: When the turbine is parked (not generating power) or idling, it still experiences environmental loads. The IEC 61400-3 standard specifics load cases for these conditions, including extreme wind gusts or waves acting on the idle turbine. These scenarios are essential for ensuring that the turbine structure remains safe, even when not in operation.

- *Service and maintenance states:* During service or maintenance activities, turbines may be subjected to unusual loads due to the presence of additional equipment or personnel on the platform. These load cases consider the impact of lifting equipment or cranes, which can shift the centre of gravity and apply temporary loads to the structure.

By thoroughly analysing these design load cases, engineers can ensure that OWT support structures remain safe, efficient, and reliable throughout their operational lifetime. The combination of ULCs and FLCs, along with appropriate environmental load combinations, provides a comprehensive framework for designing robust OWT support structures.

5. Limit states

The design of OWT support structures must ensure safety and functionality throughout their operational life under diverse environmental and operational conditions. This is achieved by analysing various limit states, which represent conditions under which the structure can no longer perform its intended function. Limit states are classified into four main categories: ultimate limit state (ULS), fatigue limit state (FLS), serviceability limit state (SLS), and accidental limit state (ALS). These categories form the foundation for designing safe and reliable OWT support structures, as outlined in *IEC 61400-3* [7] and *DNVGL-ST-0126* [8].

5.1 Ultimate limit state (ULS)

The ULS represents the maximum load-carrying capacity of a structure before failure. ULS analysis ensures that the structure can withstand the most extreme environmental and operational loads expected over its lifetime, including high wind speeds, storm waves, and severe operating conditions.

- *Yielding and buckling*: ULS assessments evaluate whether the structure will experience yielding or buckling under the applied loads. For tubular members, such as monopiles and jacket foundations, buckling can occur under high compressive forces. The design must ensure that neither the tower nor the foundation exceeds the material's yield stress or critical buckling stress.

- *Extreme environmental conditions*: ULS assessments include the impact of extreme environmental conditions, such as the 50-year return period events, where the combination of wind, wave, and current loads is at its peak.

- *Load combinations*: ULS requires considering combined load cases, including simultaneous wind, wave, and current forces. The *DNVGL-ST-0126* standard [8] provides guidelines for calculating these combined loads and determining the structural response under simultaneous extreme conditions.

5.2 Fatigue limit state (FLS)

The FLS addresses the long-term effects of cyclic loading on the OWT support structures, which experience constant fluctuations in load due to variable wind speeds, waves, and rotor movements.

- *Cyclic loading*: FLS analysis considers the effects of repeated load cycles on critical components, such as the monopile and tower base. For example, the tower experiences cyclic stresses from rotor thrust and wind-induced oscillations. These repeated stresses can cause micro-cracks that grow over time, leading to fatigue failure if not properly accounted for in the design.

- *S-N curves and cumulative damage*: FLS assessments normally use S-N curves to model the relationship between stress range and the number of cycles to failure for different materials. The Palmgren-Miner rule is typically used to calculate cumulative damage by summing the damage from each load cycle. If the cumulative damage exceeds a critical threshold, fatigue failure is expected. The design must ensure that the structure's fatigue life exceeds the expected operational life of the wind turbine.

- *Fatigue in welded joints*: Welded connections are particularly vulnerable to fatigue due to stress concentrations at the welds. The design must account for potential fatigue failure in welded joints by specifying appropriate weld quality and post-weld treatment techniques. For instance, post-weld grinding and peening can improve fatigue performance.

5.3 Serviceability limit state (SLS)

The SLS ensures that the structure remains functional and within acceptable performance criteria under normal operating conditions. Unlike ULS and FLS, which focus on failure and long-term degradation, SLS concerns operational tolerances, including deformations, vibrations, and deflections, which affect the turbine's performance and longevity.

- *Deformation and deflection*: SLS assessments check that deflections in the support structure, such as tower tilt and bending, remain within acceptable limits. Excessive deflections can affect the rotor-nacelle alignment, leading to inefficiencies in power generation and increased wear on mechanical components. The design must ensure that deflections under normal operational loads remain within specified tolerances.

- *Vibration control*: OWT support structures are subject to dynamic loading from wind and waves, which can induce vibrations in the tower and foundation. Resonant vibrations, in particular, can lead to significant structural damage if the natural frequency of the turbine aligns with the frequency of external loads. SLS analysis ensures that the turbine's natural frequency is sufficiently separated from excitation frequencies to prevent resonance.

- *Service life considerations*: The structure must maintain its serviceability over its entire design life, which is typically 20–25 years for OWTs. SLS assessments also include considerations for operational phases such as maintenance and repairs, ensuring that temporary loads during these phases do not affect the turbine's overall performance.

5.4 Accidental limit state (ALS)

The ALS ensures that the structure can withstand unforeseen, extreme events without catastrophic failure. These include scenarios such as ship collisions, dropped objects during maintenance, and extreme environmental events beyond the 50-year return period.

- *Ship collision*: OWTs are often located in busy shipping lanes, making collision risks a critical consideration. ALS assessments evaluate the potential impact forces from ship collisions and ensure that the support structure can absorb these forces without collapsing. In the case of monopile foundations, the impact energy can be dissipated through plastic deformation, but the overall integrity of the structure must be maintained.

- *Dropped objects*: During maintenance operations, there is a risk of heavy objects being dropped onto the structure. ALS analysis evaluates the potential damage from such incidents and ensures that critical components, such as the tower and foundation, remain safe under these conditions.

- *Fire and explosion*: Although rare, fire or explosion hazards must be considered, especially for floating wind turbines or those with adjacent substation platforms. ALS assessments ensure that the structure can resist these accidental loads long enough to prevent loss of life or environmental harm.

5.5 Design optimisation across limit states

To ensure long-term performance, OWT support structures must balance the requirements of ULS, FLS, SLS, and ALS. This is often achieved through optimisation techniques that minimise material usage while meeting the safety criteria for each limit state.

6. Structural modelling

Structural modelling is a critical component in the design of OWT support structures. It provides engineers with tools to predict how the structure will respond under various operational and environmental conditions. The accuracy and reliability of the model are essential in ensuring that the design can withstand the complex interactions between wind, waves, currents, and other forces. Structural models for OWTs can be broadly categorised into 1D beam models and 3D finite element analysis (FEA) models. Each type has its specific applications, advantages, and limitations.

6.1 One-dimensional (1D) beam models

The 1D beam model is a widely used, computationally efficient method for analysing the global dynamic behaviour of OWT support structures. In this approach, the tower and support structures are modelled as a series of elastic beam elements, typically using Euler-Bernoulli or Timoshenko beam theory. The beam elements represent the support structure as a slender body subjected to bending, shear, and axial loads.

- *Applications of 1D beam models*: These models are widely adopted in commercial wind turbine design software, such as GH-Bladed [11] and OpenFAST [12], for global dynamic analysis. They can predict the natural frequencies, overall deflections, and load distributions of the support structure under various operational conditions. For example, monopiles and jackets can be effectively modelled as beam elements, allowing for rapid iteration during the design process.

- *Advantages*: 1D beam models offer a good balance between computational efficiency and accuracy for global dynamic behaviour. These models are particularly useful during the early design stages, where quick assessments of the structure's overall performance are required. They are also well suited for load case simulations, including fatigue and ultimate limit states.

- *Drawbacks*: While 1D models capture the global dynamic response well, they struggle to accurately predict local stress concentrations and deformations, particularly at critical regions such as joints, welds, and transitions between structural elements. Therefore, these models are often supplemented by more detailed analyses using 3D FEA.

6.2 Three-dimensional (3D) finite element analysis (FEA) models

For a more detailed structural analysis, 3D FEA models are employed. These models break the structure into a mesh of small elements (such as brick or shell elements) that allow for a detailed examination of the stress distribution, deformations, and local structural responses. FEA models can incorporate complex geometries and material behaviours, making them indispensable for identifying areas prone to stress concentrations, such as the tower base, monopile-soil interface, and welded joints.

- *Applications of 3D FEA models*: 3D FEA models are particularly useful for analysing local effects that are not captured by 1D beam models. For example, when designing jacket foundations, 3D models can simulate the stress distribution across individual joints and tubular elements, providing insights into the risk of fatigue failure. Additionally, these models are essential for assessing how local defects, such as weld cracks or corrosion pits, might affect the structure's overall integrity.

- *Advantages*: The primary advantage of 3D FEA models is their accuracy in capturing detailed structural responses, including local deformations and stress concentrations. These models are essential for verifying the structure's performance under complex loading conditions, such as combined wind, wave, and current forces, as well as accidental load cases like ship collisions or dropped objects during maintenance.

- *Drawbacks*: The main drawback of 3D FEA is its computational intensity. These models require significantly more time and computational resources compared to 1D beam models. Furthermore, 3D FEA models can be sensitive to the quality of the mesh and the assumptions made regarding boundary conditions, which must be carefully calibrated to ensure accurate results.

7. Optimisation algorithms

The design of OWT support structures involves numerous constraints and variables, making it a complex optimisation problem. Structural optimisation seeks to minimise material use, cost, and environmental impact while ensuring that the design meets the necessary safety, serviceability, and durability requirements under extreme operational and environmental conditions. The complexity of OWT support structure design necessitates the use of advanced optimisation algorithms to efficiently explore the vast solution space and find optimal designs that balance multiple competing objectives. Commonly used algorithms include gradient-based methods, Simulated Annealing, Genetic Algorithms (GAs), Particle Swarm Optimisation, and Multi-Objective Optimisation.

7.1 Gradient-based optimisation algorithms

Gradient-based methods [13] are some of the most traditional and widely used approaches in structural optimisation. These algorithms leverage the gradient (or derivative) of the objective function with respect to the design variables to guide the search towards optimal solutions. In the context of wind turbine structures, gradient-based methods can be applied to problems with well-defined, differentiable objective functions, such as minimising material usage while satisfying constraints related to stress, deflection, and natural frequency.

- *Advantages*: Gradient-based methods are computationally efficient, making them suitable for problems where the objective function is smooth and continuous. They converge quickly and are ideal for optimising simpler structures where the number of variables and constraints is limited.

- *Limitations*: The primary limitation of gradient-based methods is that they rely on the objective function being differentiable. For problems with non-linear constraints, discontinuities, or complex design spaces, such as those involving soil-structure interaction or fatigue failure, gradient-based methods may struggle to find the global optimum and can become trapped in local optima. Additionally, they require a good initial guess to ensure convergence towards the global optimum.

7.2 Simulated annealing (SA)

SA [14] is another popular optimisation technique, inspired by the annealing process in metallurgy, where a material is heated and then slowly cooled to remove defects and find a low-energy configuration. SA is particularly useful for solving optimisation problems with complex, non-linear, or non-convex design spaces. In SA, a candidate solution is perturbed, and the change in objective function is evaluated. If the new solution is better, it is accepted; if it is worse, it may still be accepted with a probability that decreases as the algorithm progresses, allowing SA to escape local minima early in the search process.

- Applications in OWT design: SA can be applied to optimise complex design problems, such as the placement and sizing of structural elements in jacket foundations or the configuration of floating platform mooring systems. The algorithm's ability to escape local minima makes it particularly useful for optimising designs where traditional methods fail to find the global optimum.

- Advantages: SA is robust and can handle both discrete and continuous variables. Its stochastic nature allows it to explore a wide range of potential solutions, and its decreasing acceptance probability ensures that the algorithm converges to a global optimum over time. SA is also relatively simple to implement and can be combined with other optimisation techniques for enhanced performance.

- Limitations: The main limitation of SA is its relatively slow convergence speed, particularly for large-scale optimisation problems. It may require many iterations to find an optimal solution, and the algorithm's performance is sensitive to the choice of cooling schedule and initial temperature.

7.3 Genetic algorithms (GAs)

GAs [4] are among the most popular metaheuristic optimisation techniques used in the design of wind turbine structures [15–17]. Inspired by the process of natural selection, GAs operate by iteratively evolving a population of candidate solutions through selection, crossover, and mutation processes. The algorithm evaluates the fitness of each individual solution based on an objective function, such as minimising material costs or maximising fatigue resistance.

- *Applications in OWT design*: GAs are well suited for complex design problems that involve non-linear constraints and discrete variables. In OWT design, GAs have been used to optimise the geometry and material distribution of monopiles, jackets, and floating platforms, where the design space is too large and complex for gradient-based methods to handle effectively. For example, a GA can explore different configurations of pile diameters, wall thicknesses, and material types to minimise cost while satisfying performance constraints like ultimate load capacity and fatigue life.

- *Advantages*: One of the key strengths of GAs is their ability to escape local optima and explore large, non-convex design spaces. This makes them particularly useful for solving multi-objective optimisation problems, where several conflicting goals, such as minimising cost and maximising structural integrity, need to be balanced. GAs do not require the objective function to be differentiable, making them ideal for solving real-world engineering problems involving complex load cases, soil-structure interaction, and material non-linearity.

- *Limitations*: While GAs are powerful for exploring large solution spaces, they can be computationally expensive, especially when applied to high-fidelity models such as 3D finite element analyses. Convergence towards the global optimum can be slow, and the algorithm's performance is sensitive to the choice of parameters, such as population size, crossover rate, and mutation rate. In some cases, GAs may require hybridisation with other methods to enhance convergence speed.

7.4 Particle swarm optimisation (PSO)

Another metaheuristic algorithm frequently used in structural optimisation is PSO [18], which simulates the social behaviour of organisms such as birds flocking or fish schooling. In PSO, a swarm of particles (candidate solutions) moves through the design space, adjusting their position based on both their own experience and that of neighbouring particles. Each particle represents a potential solution, and its movement is guided by the best solutions found by itself and the swarm.

- *Applications in OWT design*: PSO has been successfully applied to optimise the layout and design of wind farm support structures, particularly for jacket and floating platforms. For example, PSO can be used to minimise the mass of a jacket structure while ensuring that its natural frequency avoids resonance with the wave and wind loading frequencies. PSO is also useful for optimising mooring system configurations for floating turbines, where the design variables include cable lengths, tensions, and anchor locations.

- *Advantages*: PSO is easy to implement and has fewer parameters to tune compared to GAs. It is highly effective at exploring continuous design spaces and can find near-optimal solutions relatively quickly. PSO is also less prone to becoming trapped in local optima compared to gradient-based methods, making it suitable for problems with complex, non-linear objective functions.
- *Limitations*: Like GAs, PSO can be computationally expensive, especially when applied to high-fidelity models. It may also struggle with problems involving discrete variables or highly constrained design spaces. Additionally, the convergence speed of PSO can be slower compared to other metaheuristic algorithms, particularly when dealing with multi-objective problems.

7.5 Multi-objective optimisation

Many design problems in OWT support structures involve multi-objective optimisation, where several objectives, such as cost, weight, fatigue life, and structural integrity, must be optimised simultaneously. Multi-objective algorithms, such as NSGA-II (Non-dominated Sorting Genetic Algorithm II) [19], are commonly used to generate a set of Pareto-optimal solutions, where no single solution is strictly better than another across all objectives.

- *Advantages of multi-objective optimisation*: These algorithms allow designers to explore trade-offs between competing objectives and provide decision-makers with a set of optimal solutions to choose from. For instance, in the design of jacket structures, a multi-objective algorithm might generate a set of solutions that trade-off between minimising material cost and maximising fatigue life, allowing engineers to select a solution based on project-specific priorities.
- *Limitations*: Multi-objective optimisation algorithms can be computationally demanding, particularly when applied to high-fidelity structural models. Convergence to the Pareto front can be slow, and the results may be difficult to interpret when there are many conflicting objectives.

8. Case study: Integrated design optimisation of OWT support structures

In this case study, the integrated structural optimisation of an OWT support structure was performed. The detailed description on this case study can be found in Ref. [15]. The objective of this study was to develop an optimisation model for OWT support structures that minimises mass while ensuring compliance with various design constraints, such as stress, deformation, buckling, fatigue, and vibration. The study employed a coupled FEA and GA approach, and it was applied to the National Renewable Energy Laboratory (NREL) 5 MW wind turbine [20] on an OC3 (offshore code comparison collaboration) monopile foundation [21]. The results showed a 19.8% reduction in the global mass of the support structure, demonstrating the efficiency and practicality of the proposed optimisation model.

8.1 Background and motivation

Support structures account for approximately 25% of the capital cost of an OWT. Reducing their mass and cost without compromising safety and performance is a key factor in decreasing the overall cost of offshore wind energy. Traditional design approaches for OWT support structures often optimise individual components, such as the tower or monopile, separately. However, this study adopted an integrated approach, optimising the entire support structure simultaneously. By using both FEA to simulate the structural response under various loading conditions and GA to explore the vast design space efficiently, the model aimed to find optimal designs that satisfy multiple criteria while reducing material usage and cost.

8.2 Structural modelling

The NREL 5 MW wind turbine [20] used in this study has a rated power of 5 MW, with a rotor diameter of 126 metres and a 77.6-metre-high tower. The turbine was assumed to operate in a Class I wind regime. The OC3 monopile foundation, adapted for the NREL turbine, consists of a 56-metre-long pile with a diameter of 6 metres and a thickness of 60 mm, embedded 36 metres into the seabed.

A key aspect of the optimisation model was its detailed representation of the support structure using a parametric 3D FEA model. The model included the tower, transition piece, monopile, and soil, with shell elements used for thin-walled components such as the tower and monopile, and solid elements used for the soil and grout connections. The soil-structure interaction was considered, and the constitutive model of the soil was described using the Drucker-Prager model, which is widely used for simulating soil behaviour. This detailed modelling ensured accurate predictions of stress distributions, deformations, and other structural responses critical for designing safe and efficient support structures.

8.3 Optimisation process

Figure 3 depicts the flowchart of the structural optimisation model of OWT support structures [15], which combines the parametric FEA model and the GA model. The optimisation objective was to minimise the global mass of the support structure. The design variables included the diameters and thicknesses of various segments of the tower and monopile, with a total of 13 design variables. Constraints were applied to ensure compliance with ultimate limit state (ULS), serviceability limit state (SLS), and fatigue limit state (FLS) criteria. These constraints included vibration, stress, deformation, buckling, and fatigue safety.

The GA was used to search for optimal solutions. The algorithm began with an initial population of design solutions, which were evaluated using FEA. Based on the fitness of each solution, the GA iteratively generated new populations by applying crossover and mutation operators, gradually converging towards the optimal design. The stopping criterion was set at 25 iterations or when the optimisation stabilised. More details of the optimisation model can be found in Ref. [15].

8.4 Results and discussion

The optimisation process resulted in a 19.8% reduction in the total mass of the support structure, from 924 tonnes to 742 tonnes. The largest mass savings were achieved in the monopile, which saw a reduction of 27%, while the tower's mass decreased by 10%. Interestingly, the transition piece's mass increased slightly (by 1%), indicating that the optimal solution did not involve uniformly reducing the mass of all

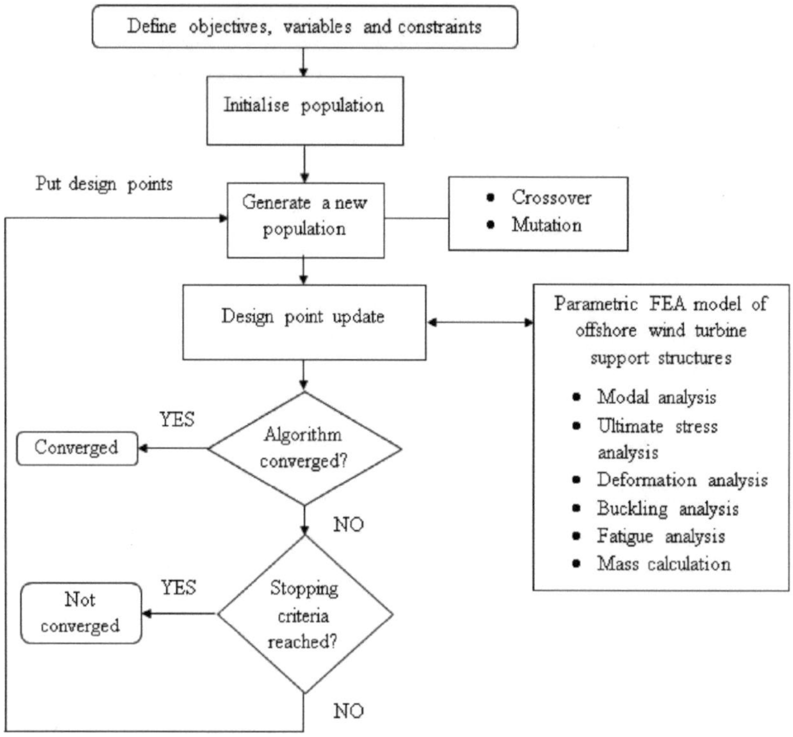

Figure 3.
Flowchart of the optimisation model [15].

components. Instead, the integrated approach allowed for trade-offs between different parts of the support structure to achieve an overall optimal design.

The design was primarily constrained by the first natural frequency and fatigue, which were the most saturated constraints during the optimisation process. The first natural frequency had to lie between the blade passing frequency and rotor-induced frequency to avoid resonance. Fatigue analysis was critical due to the cyclic nature of the loads experienced by the support structure during the turbine's operation.

8.5 Conclusions

This case study demonstrated the effectiveness of integrating FEA with GA for the optimisation of OWT support structures. By considering the entire structure as a single system and optimising it holistically, the model achieved significant material savings while meeting all design constraints. This approach not only reduces the cost of offshore wind energy but also improves the design efficiency and reliability of wind turbine support structures.

The results of this study highlight the potential of advanced optimisation techniques in reducing the cost of offshore wind energy and making it more competitive in the global energy market. Future research could explore further improvements in computational efficiency and apply this methodology to other types of support structures, such as jacket or floating foundations.

Acknowledgements

This work was supported by grant RIR38231129-3 from the UK Engineering and Physical Sciences Research Council (EPSRC) through the Innovation Launchpad Network+ Researchers in Residence Scheme. We also acknowledge the valuable collaboration and support offered by Offshore Renewable Energy Catapult.

Author details

Lin Wang[1*] and Athanasios Kolios[2]

1 Department of Mechanical, Materials and Manufacturing Engineering, University of Nottingham, Nottingham, United Kingdom

2 Department of Wind and Energy Systems, Technical University Denmark, Roskilde, Denmark

*Address all correspondence to: lin.wang1@nottingham.ac.uk

References

[1] GWEC. Global Wind Report 2023. Brussels, Belgium: Global Wind Energy Council; 2023

[2] Martinez-Luengo M, Kolios A, Wang L. Parametric FEA modelling of offshore wind turbine support structures: Towards scaling-up and CAPEX reduction. International Journal of Marine Energy. 2017;**19**:16-31

[3] Wang L, Quant R, Kolios A. Fluid structure interaction modelling of horizontal-axis wind turbine blades based on CFD and FEA. Journal of Wind Engineering and Industrial Aerodynamics. 2016;**158**:11-25

[4] Mirjalili S. Genetic algorithm. In: Evolutionary Algorithms and Neural Networks: Theory and Applications. Cham, Switzerland: Springer; 2019. pp. 43-55

[5] Wang Y, Zhang H, Liang F, Wang C, Yuan Z. Seismic response of offshore tetrapod piled jacket foundations subjected to environmental loads in soft-over-stiff clay deposit. Computers and Geotechnics. 2024;**173**:106547

[6] Liu Y, Li S, Yi Q, Chen D. Developments in semi-submersible floating foundations supporting wind turbines: A comprehensive review. Renewable and Sustainable Energy Reviews. 2016;**60**:433-449

[7] Quarton D. An International Design Standard for Offshore Wind Turbines: IEC 61400–3. Bristol, UK: Garrad Hassan and Partners, Ltd; 2019

[8] DNVGL. DNVGL-ST-0126 Support Structures for Wind Turbines. Høvik, Norway: DNVGL; 2021

[9] Chehouri A, Younes R, Ilinca A, Perron J. Review of performance optimization techniques applied to wind turbines. Applied Energy. 2015;**142**: 361-388

[10] DNVGL. DNVGL-ST-0126: Support Structures for Wind Turbines. Høvik, Norway: DNVGL; 2016

[11] DNVGL. Bladed Theory Manual. Høvik, Norway: DNVGL; 2021

[12] Jonkman J. National Renewable Energy Laboratory technical report. In: OpenFAST User's Guide. Golden, CO: National Renewable Energy Laboratory; 2021

[13] Chew K-H, Tai K, Ng E, Muskulus M. Analytical gradient-based optimization of offshore wind turbine substructures under fatigue and extreme loads. Marine Structures. 2016;**47**:23-41

[14] Nikolaev AG, Jacobson SH. Simulated annealing. In: Handbook of Metaheuristics. Cham, Switzerland: Springer; 2010. pp. 1-39

[15] Gentils T, Wang L, Kolios A. Integrated structural optimisation of offshore wind turbine support structures based on finite element analysis and genetic algorithm. Applied Energy. 2017; **199**:187-204

[16] Wang L, Kolios A, Nishino T, Delafin P-L, Bird T. Structural optimisation of vertical-axis wind turbine composite blades based on finite element analysis and genetic algorithm. Composite Structures. 2016;**153**:123-138

[17] Al-Sanad S, Parol J, Wang L, Kolios A. Structural optimisation framework for onshore wind turbine towers considering multiple design constraints. International Journal of

Sustainable Energy. 2022;**41**(5): 469-491

[18] Wang D, Tan D, Liu L. Particle swarm optimization algorithm: An overview. Soft Computing. 2018;**22**(2): 387-408

[19] Deb K, Agrawal S, Pratap A, Meyarivan T. A fast elitist non-dominated sorting genetic algorithm for multi-objective optimization: NSGA-II. In: Parallel Problem Solving from Nature PPSN VI: 6th International Conference; 18–20 September 2000; Paris, France. Springer; pp. 849-858

[20] Jonkman J. Definition of a 5-MW Reference Wind Turbine for Offshore System Development. Golden, CO: National Renewable Energy Laboratory; 2009

[21] Jonkman J, Musial W. Offshore Code Comparison Collaboration (OC3) for IEA Wind Task 23 Offshore Wind Technology and Deployment. Golden, CO: National Renewable Energy Lab (NREL); 2010

Section 2

Engineering Modeling and Applications

Chapter 4

Manufacturing and Testing a Small Wind Turbine Blade Using 3D Printing

Natasha Martins Rodrigues de Jesus, Thais Santos Castro, Teófilo Miguel de Souza, Raphaela Carvalho Machado, José M. Balthazar, Clivaldo de Oliveira, Jeferson Jose de Lima and Angelo M. Tusset

Abstract

The current critical global climate situation, which is due to greenhouse gas emissions, has led society to look for ways to preserve the environment. One of these is the use of renewable energies, including wind power. This energy uses the wind to operate. Based on this, this work seeks to design and produce wind blades using polymeric materials in an additive manufacturing process, in this case, with a 3D printer. The materials used were ABS, PETG, PLA and Tritan. The shapes developed were straight, curved and curved tip for the Tritan polymeric material, while for the other materials, the shape designed was straight. Once manufactured, the blades were subjected to performance tests with a wind turbine coupled to an alternator consisting of a tripod turbine. During the experiments, the construction materials, the shapes and the angle of attack as a function of the linear wind speed generated by a wind tunnel were changed so that the analysis was consistent. The results obtained were compared to identify the best combinations of the parameters analyzed. In the end, it was concluded that the blades printed with the Tritan polymeric material could be applied to various wind speed conditions—low, medium and high—as well as having good mechanical properties and the best performance in their manufacturing process, making them the best of those studied.

Keywords: three-piece turbine, wind energy, Catia V5, 3D CAD modeling, 3D printing

1. Introduction

Due to the increase in consumption and the world's dependence on energy generated from non-renewable resources, the concept of sustainable energy is highlighted in today's society and the need for alternative energy systems [1].

One of the world's biggest challenges and concerns is the emission of greenhouse gases due to climate change, especially the damage caused to the ozone layer which, as a result, increases the ambient temperature. In this respect, Brazil showed low emission rates in 2018 and emitted around 85% less than China, 76% less than the United States and 69% less than the European Union to generate each MWh, according to the 2021 Electricity Statistical Yearbook [2].

In view of this, according to the UN [3], the importance of environmental preservation in Brazil includes viable, sustainable, renewable and modern energy for all in its sustainability goals. To this end, energy sources involve the use of cleaner and more renewable elements [4]. According to Carr and Thomson [5], energy efficiency is the most effective way of reducing costs and local and global environmental impacts at the same time. The use of renewable energy sources has major benefits, such as reducing atmospheric greenhouse gas emissions, providing a means of diversifying energy and fuel supplies, and reducing energy costs for consumers. Among the renewable energy sources, wind energy, solar energy and biomass energy are currently the most studied. Wind energy accounts for 13.4% of the Brazilian matrix, followed by solar energy with 19.8% and biomass with 7.5% [6].

According to the Global Wind Energy Council (GWEC), the country has been growing consistently and was the country with the third largest number of onshore wind energy installations in 2023 with 4.8 GW, with a decrease in 2024, but still remains in sixth place in the ranking with 30.45 GW, as shown in **Figure 1** according to GWEC [7].

Wind energy is a renewable and sustainable energy, one of the most advanced in terms of technology and application, because as well as not emitting polluting gases, it uses an inexhaustible source of fuel, the wind [8]. It is currently one of the options responsible for increasing the supply of new energy sources and has a low cost compared to other renewable energy sources [9].

In addition to wind technology, additive manufacturing has shown an exponential and increasingly specific development [10]. The current form of manufacturing is made up of a set of additive manufacturing processes that enable the production of prototypes, parts and devices through the extrusion and deposition of various materials, such as polymer, in the desired geometry [11].

With this in mind and understanding the importance of the growth of the additive manufacturing sector for industries, the aim of this article is to present an innovative process for manufacturing wind blades for small horizontal wind turbine systems, using polymeric materials and the 3D printing process to manufacture them.

GWEC 2023 RANKING - TOTAL INSTALLED CAPACITY ONSHORE		
Position	Country	Total installed capacity onshose (GW)
1	China	403.33
2	USA	150.43
3	German	61.14
4	India	44.74
5	Spain	30.56
6	Brazil	30.45
7	France	22.00
8	Canada	16.99
9	Sweden	16.25
10	United Kingdom	14.87

Figure 1.
GWEC 2024 ranking shows the installed capacity of onshore wind power.

2. Materials and methods

The research is of a basic and applied nature, with exploratory objectives and a quantitative approach using the experiment method. This is an approach in which the study, by delving into associations or correlations, analyzes and makes inferences about the reasons for causes and how these causes are established; this method is aimed at exploratory data sets rather than individual data [11]. To carry out the experiments, two test groups were defined: the first kept the straight blade format and varied the ABS, PETG and PLA materials. The second set kept the Tritan material and varied the straight, curved and curved tip shapes. The curved tip blade was made in two angles by reversing the direction of the blade's trailing edge, and was given the names curved tip 1 and curved tip 2. The wind blades were made from polymeric materials. The dimensions of the blades are: The straight blade is 570 mm long, 114.2 mm wide at the root, 41.3 mm wide at the tip, thicker by 17 mm and thinner by 5 mm tip, thicker by 17 mm and thinner by 5.8 mm. The curved blade is 570 mm long, 114.5 mm wide at the root, 22.2 mm wide at the tip, thicker by 17 mm and thinner by 4 mm tip width, thicker by 17 mm and thinner by 4 mm. The curved tip shovel has a length of 570 mm, a root width of 80.8 mm, a tip width of 15.5 mm, a greatest thickness of 17.2 mm and a least thickness of 3 mm tip width, thicker by 17.2 mm and thinner by 3 mm.

For the tests, a Delco-Arno alternator was used to generate 400 W/12 VDC/35 A/1000 rpm coupled to the Primus Air X Marine small wind turbine [12]. The blades fixed to it would have the same dimensions as the blades designed for the model used in this work. The construction of a wind turbine requires the following components: the blades, the rotor, the nacelle and the tower [13]. Wind turbines can be classified into three groups due to the variety of sizes available; this division is based on the power of each one [14]:

- Small: Power from 0.1 to 100 kW;
- Medium-sized: Power from 101 to 300 kW;
- Large: Power above 300 kW.

There are also two basic types of wind turbine: vertical axis and horizontal axis [15]. This division of the two groups is defined according to the orientation of the rotor's axis of rotation [16]. In this work, the focus is on the horizontal axis model.

A wind tunnel was used to simulate the linear wind speed, as well as measuring devices such as a digital anemometer, digital tachometer, DC ammeter, digital inclinometer, DC multimeter and digital thermometer. **Figure 2** shows the schematic drawing of the test.

The study established the following requirements for the choice of polymeric materials: firm structure for greater mechanical strength to withstand aerodynamic forces and fatigue caused by cyclical and repetitive movements. Resistance to creep, since wind turbine blades are normally exposed to various climatic conditions, and resistance to bending, since depending on the speed reached by the blade, it can increase turbulence with the flow of wind and increase the stresses exerted on the part (which can cause a torsion angle at the tip of the blade, reducing the diameter of the turbine and eventually causing the structure to break); surface finish for better performance of the blades and care with the roughness of their surface, since the lower its value, the better their efficiency.

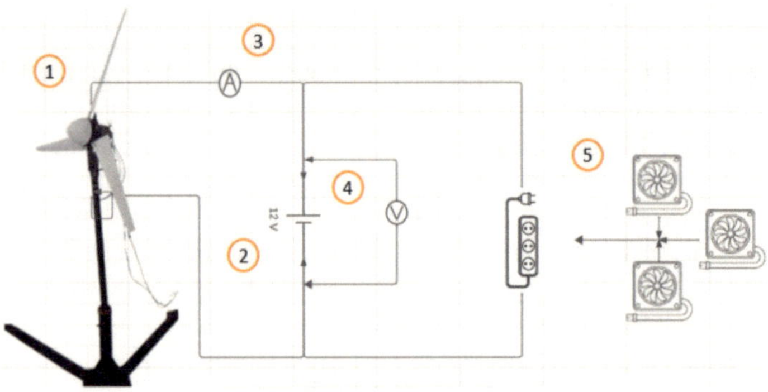

Figure 2.
Schematic drawing of the test carried out (1) wind turbine; (2) battery; (3) ammeter; (4) multimeter and (5) coolers.

The description of the electrical diagram for the tripe turbine's power generation is shown in **Figure 2**, begins with item (1) a 1.20-meter tripod support that holds up the wind turbine structure coupled with an alternator connected to a cable in the battery, for charging, (2) battery for storing the charge and for testing the charging and discharging of the coolers, (3) ammeter for DC voltage measurements at the output of the wind turbine connected to the battery, (4) multimeter used as a voltmeter, for DC voltage measurements in the battery, (5) set of 21 W coolers connected to a cable in the socket. DC voltage measurements on the battery and (6) set of 21 W coolers connected to a cable in the socket that connected to the socket strip, to carry out the discharge in the experiment.

The test consisted of placing the wind blades under controlled wind conditions obtained in the laboratory. The angles of attack used were 5°, 10°, 15°, 20° and 25°, established according to the NACA 2412 aerodynamic profiles [17].

Figure 3 shows a model with a straight blade shape.

Figure 3.
Small wind turbine used in the study.

3. Results

During the course of the experiment there was a proportional increase between the linear wind speed and the frequency of the wind tunnel fan. As it is a controlled environment, the variations in the environment are minimal and it can be used for all the experiments, with a controlled temperature of 25°C in a closed laboratory and without external interference. These environmental conditions will be used for all sets of blades.

The result is presented as a graph between the blades' revolutions per minute and the power generated by the turbine. The maximum load to which the wind turbines were subjected was 21 W/12VDC. It was necessary to establish a maximum load so that the generator could supply the battery and not discharge it.

The graph in **Figure 4** shows the revolutions as a function of the angles of attack for linear speed for the ABS, PETG and PLA material for the straight wind blade format.

It can also be seen in **Figure 4** that the blade with the best performance was the PLA material blade for angles of attack of 10°, 15°, 20° and 25°, since for the angle of attack of 5° none of the blades overcame the inertia of the movement to start rotating at low linear wind speed.

Figure 5 shows that it was easier for the blades to overcome the inertia of the movement to start rotating, reaching higher rpm values than the low wind speed condition, the condition for this phase of the experiment being an average linear wind speed of 6.6 m/s. It can be seen here that the blade with the best performance for the conditions of varying the angle of attack by 10°, 15°, 20° and 25° was the PLA polymer blade.

Figure 6 shows the graph with linear wind speed conditions of 11.7 m/s, considered a high speed. For this condition, the blade with the best performance was the PLA blade.

The experiments carried out on the straight blades were the same as those applied to the Tritan material, with the only changes being to the geometries of the wind blades, which were straight, curved and curved tip 1 and curved tip 2. The blades, curved tip 1 and 2, are so named because the experiments were carried out with the

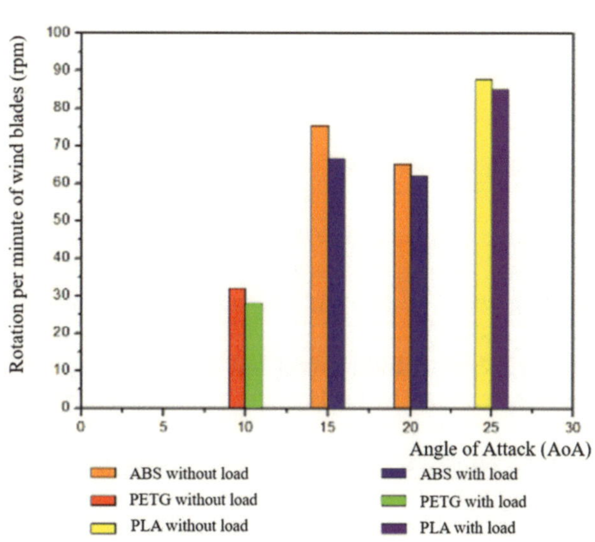

Figure 4.
Rotations as a function of varying angles of attack with a linear wind speed of 3.6 m/s.

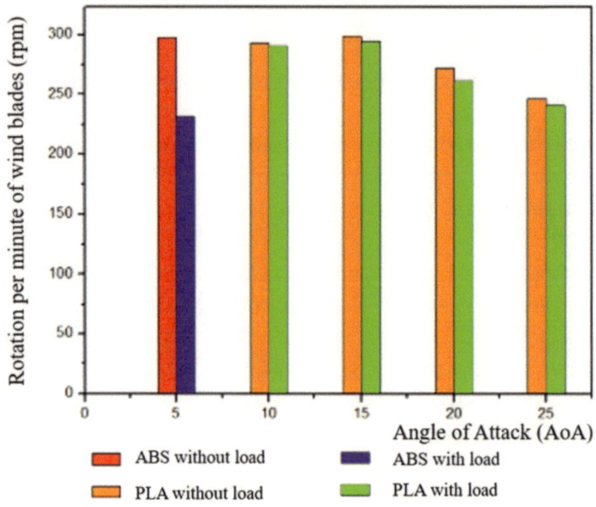

Figure 5.
Rotations as a function of varying angles of attack with a linear wind speed of 6.6 m/s.

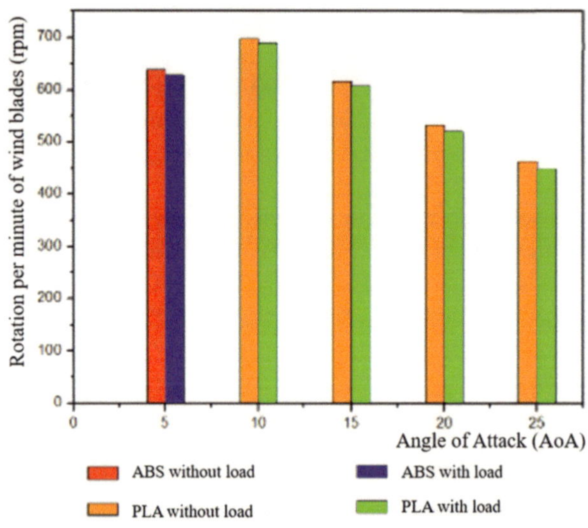

Figure 6.
Rotations as a function of varying angles of attack for a linear wind speed of 11.7 m/s.

blade's trailing edge reversed. The load and alternator used in these experiments were the same as in the previous items.

The graph in **Figure 7** shows conditions of low linear wind speed for its respective speed of 3.6 m/s. This shows that the blade assembly that performed best was the curved one.

The graph shown in **Figure 8** shows average linear wind speed conditions for the respective speed of 6.6 m/s. In this part of the experiment, it can also be seen that the blade assembly that performed best was the curved shape for the conditions imposed at this stage.

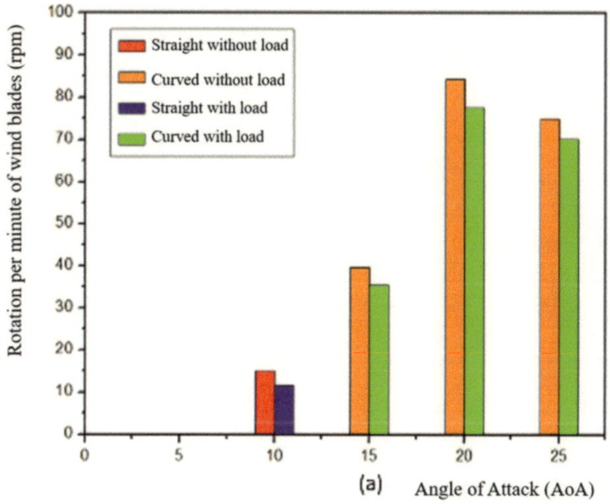

Figure 7.
Rotations as a function of varying angles of attack for the Tritan blades at a linear wind speed of 3.6 m/s.

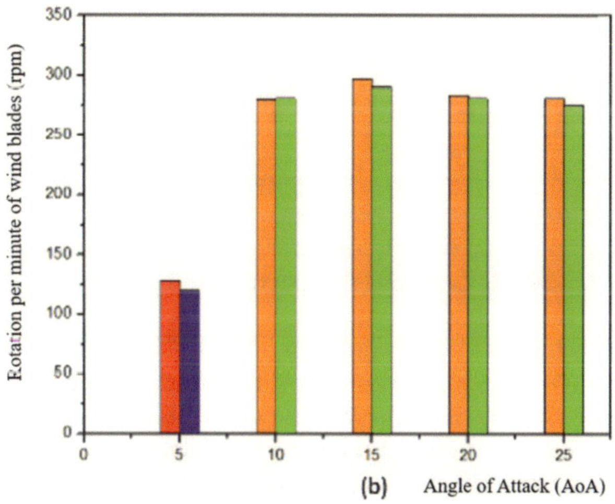

Figure 8.
Rotations as a function of angle of attack for the Tritan blades at a linear wind speed of 6.6 m/s.

Finally, the graph in **Figure 9** shows high linear wind speed conditions for speeds of 11.7 m/s. Likewise, the blade assembly that performed best was the curved shape.

The first analysis of the graphs in **Figures 4–6** shows that the blades with the best performance were the straight PLA blades, and the second analysis of the graphs for the wind blades printed on Tritan polymeric material in **Figures 7–9** shows that the blade set with the curved geometry had the best performance. It was also possible to see that there is a more suitable model of material and manufacturing profile for the different wind speeds. For lower speeds, the ideal blade, according to the test results, is the curved Tritan wind blade, and for higher wind speeds, it is the straight PLA wind blade.

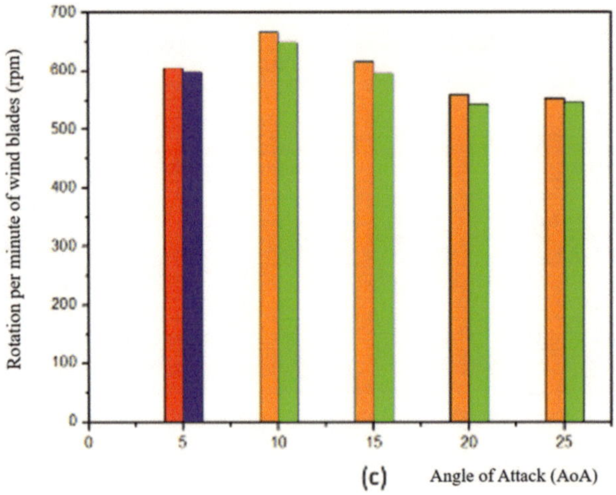

Figure 9.
Rotations as a function of angle of attack for the Tritan blades at a linear wind speed of 11.7 m/s.

4. Conclusions

The conclusion of this project is based on the manufacture and testing of wind blades in polymeric materials using 3D printing.

With regard to the 3D printing process, it was possible to conclude that the material that was easiest to print was the straight PLA blade format because the printing process did not require higher machine temperatures, and it was the piece that had the best finish with the shortest printing time.

For all the wind speed conditions, low, medium and high, to which the Tritan wind blades were subjected, the shape that performed best was the curved shape with angle variation.

It concludes that there is a more suitable model of blade material and profile for each wind speed range and that this range covers both small and large volumes.

Finally, future work could include painting the wind blades to improve the finish of the parts, studies on the surface roughness of the blades and the surface roughness of the blades and analyzing whether there is an influence and how the parts behave, improving the design of the blades (modeling in which the parts have a wider area and with more printing area), developing the printing of the blades that performed the best performance.

Author details

Natasha Martins Rodrigues de Jesus[1], Thais Santos Castro[1], Teófilo Miguel de Souza[1], Raphaela Carvalho Machado[1], José M. Balthazar[2,3]*, Clivaldo de Oliveira[4], Jeferson Jose de Lima[3] and Angelo M. Tusset[3]

1 School of Engineering, São Paulo State University (UNESP), Guaratinguetá, SP, Brazil

2 São Paulo State University (UNESP), Bauru, SP, Brazil

3 Federal University of Technology–Parana, Ponta Grossa, PR, Brazil

4 Faculty of Engineering, Federal University of Grande Dourados, Dourados, MS, Brazil

*Address all correspondence to: jmbaltha@gmail.com

References

[1] De Almeida ECV, Potentiality of photovoltaic solar energy in the northeastern semi-arid and its relationship to sustainable development [83f Dissertation Graduate Program in Regional Development – PPGDR]. State University of Paraíba; 2021. Available fromt: https://bdtd.ibict.br/vufind/Record/UEPB_33b976870d7d2ce4a632c1033a9b5e9d?print=1 [Accessed: January 17, 2025]

[2] Electricity Statistics. Statistical Yearbook of Electricity 2021. Brasília: EPE; 2021

[3] United Nations Organization. United Nations in Brazil: Investment in Renewable Energy Surpasses that of Fossil Fuels in 2018 in the World. Brasília: UN; 2018. Available from: https://brasil.un.org/pt-br/83454-investimento-em-energias-renováveis-supera-o-de-combustíveis-fósseis-em-2018-no-mundo [Accessed: August 21, 2023]

[4] Lima WCS, Hardt LPA, Hardt C, Hardt M. Interactions between the environment, sustainable development and the circular economy: Climate change and resilient cities: From challenges to solutions. Journal of Law, Economics and Sustainable Development. 2022;**8**(1):1-67

[5] Carr D, Thomson M. Non-Technical Electricity Losses. MDPI; 2022. Available from: https://www.mdpi.com/1996-1073/15/6/2218 [Accessed: January 17, 2025]

[6] ABSOLAR. Brazilian Electricity Matrix 2024. 71st ed. Bela Vista: Brazilian Photovoltaic Solar Energy Association; 2024. p. 1. Available from: https://www.absolar.org.br/mercado/infografico/ [Accessed: September 25, 2024]

[7] GWEC. Global Wind Report 2024. 1st ed. Brussels: Global Wind Report Council. p. 168. Available from: https://gwec.net/wp-content/uploads/2024/04/GWR-2024_digital-version_final-1.pdf [Accessed: September 25, 2024]

[8] Freire AÍ, Fontgalland IL. Prospects and economic challenges of wind power generation in the northeast region of Brazil. Research, Society and Development. 2022;**11**(1):e58911125429

[9] De Azevedo TSP. Sizing a Permanent Magnet Synchronous Generator for Wind Energy. Double Degree master's with UTFPR. Federal Technological University of Paraná; 2022. Available from: https://bibliotecadigital.ipb.pt/bitstream/10198/25750/1/Pedro%20Azevedo.pdf [Accessed: January 17, 2025]

[10] Cabreira V. Evaluation of 3D printing processing parameters on the properties of poly lactic acid. Dissertation (master's degree)]. Porto Alegre: Federal University of Rio Grande do Sul; 2018

[11] Gonzaga GL. Biodegradation behavior of natural and pigmented PLA samples obtained by 3D printing. 2021;**32**(2):83. Available from: https://repositorio.fei.edu.br/items/b5295815-bc8e-4925-b5b6-917996b8198/full [Accessed: January 17, 2025]

[12] Sampaio TB. Research Methodology. 1st ed. Federal University of Santa Maria; 2022. pp. 1-60

[13] Primus Wind Power. Owner's Manual Installation. Operation: Maintenance: Air X Marine. Lakewood: Primus Wind Power; 2023

[14] Silveira ADS. Technical and Economic Analysis for the

Implementation of a Residential Wind Generator [Monograph (Degree in Energy Systems Technology)]. Federal Institute of Education, Science and Technology of Santa Catarina; 2021. Available from: https://repositorio.ifsc.edu.br/handle/123456789/2207 [Accessed: January 17, 2025]

[15] Santos FA et al. Low velocity impact response of 3D printed structures formed by cellular metamaterials and stiffening plates: PLA vs. PETg. Composite Structures. 2021. Available from: https://linkinghub.elsevier.com/retrieve/pii/S0263822320330543;**256**:113128 [Accessed: March 30, 2022]

[16] Baracat PAA. Low-Cost Computational Methodology for Analyzing Vertical Axis Wind Turbines with Straight Blades [Dissertation (Master's in Mechanical Engineering)]. State University of Campinas, Faculty of Mechanical Engineering; 2021. Available from: https://repositorio.unicamp.br/acervo/detalhe/1165190 [Accessed: January 17, 2025]

[17] Airfoil Tools. NACA 1412 (naca1412-il). 2023. Available from: http://airfoiltools.com/airfoil/details?airfoil=naca1412-il [Accessed: March 30, 2022]

Chapter 5

Improved Monopile-Soil Interaction Modeling through In Situ Monitoring

Bruno Stuyts, Wout Weijtjens, Carlos Sastra Jurado and Christof Devriendt

Abstract

The performance of offshore wind turbines is strongly influenced by the structural dynamics of the overall support structure. In situ measurements of the natural frequency of monopile-supported wind turbine structures have highlighted a mismatch between the as-designed and as-built natural frequencies of these structures. This design inaccuracy can be attributed to the lateral stiffness provided by the subsoil. Accurate characterization of the lateral support provided by the soil is further complicated by the non-linear and cyclic behavior of soil. Recent research on monopile soil-structure interaction has focused on onshore field testing and numerical modeling but there are limited investigations the in situ behavior of full-scale offshore wind turbines. In this contribution, in situ monitoring data from two Belgian offshore wind farms is used to investigate the applicability of the recently developed pile-soil interaction guidance to full-scale offshore structures and to suggest improvements where necessary. The research focuses on back-analysis of bending moments during quasi-static thrust loading. Strain measurements along the length axis of the monopile allow direct measurement of the bending moment absorption below the mudline. The monitoring data reveals an important contribution of the scour protection system installed around the foundation, which can be accounted for in the pile-soil interaction calculations.

Keywords: monitoring, geotechnical, monopile, bending moment, natural frequency

1. Introduction

In this contribution, the geotechnical modeling of lateral pile-soil interaction for offshore wind turbine monopiles is explored. Monopile foundations are the most widely used foundation type for existing offshore wind farm developments with 81% of all installed offshore wind turbine adopting this foundation type [1]. By leveraging in situ monitoring data from two offshore wind farms on the Belgian continental shelf, the research aims to validate geotechnical design assumptions against measurements taken on operational full-scale offshore structures. The work seeks to establish a holistic framework for pile-soil interaction modeling that is validated by field data.

The performance of geotechnical design methods can be evaluated by comparing the as-designed bending moments along the monopiles against bending moments inferred from strain measurements. The bending strain in the monopile can be measured using strain gauges (SG) and fiber Bragg gratings (FBG) [2].

In situ strain measurements were collected at a selected offshore wind turbine structure in the Belgian North Sea, allowing the applicability of geotechnical design methods to be checked. The different types of sensors and their position on the monopile-supported wind turbine structure are shown in **Figure 1**.

The chapter is divided into three main parts, each addressing different aspects of geotechnical pile-soil interaction modeling:

1. State-of-art in pile-soil interaction modeling: Section 2 provides an overview of current methodologies for modeling pile-soil interaction, emphasizing non-linear beam column models used in monopile geotechnical design. It highlights the need for careful selection of soil mechanical parameters.

2. Improvements to pile-soil interaction modeling: Section 3 focuses on techniques for better characterization of lateral pile-soil interaction specifically for North Sea wind farm sites. This includes a discussion on the improvements made with the PISA method [3] and guidance for the derivation of reliable soil stiffness profiles from Cone Penetratin Test (CPT) data. The development of a novel stress-dependent correlation between CPT data and small-strain stiffness is discussed, which is a prerequisite for further modeling of the effect of the scour protection layer on the pile-soil interaction.

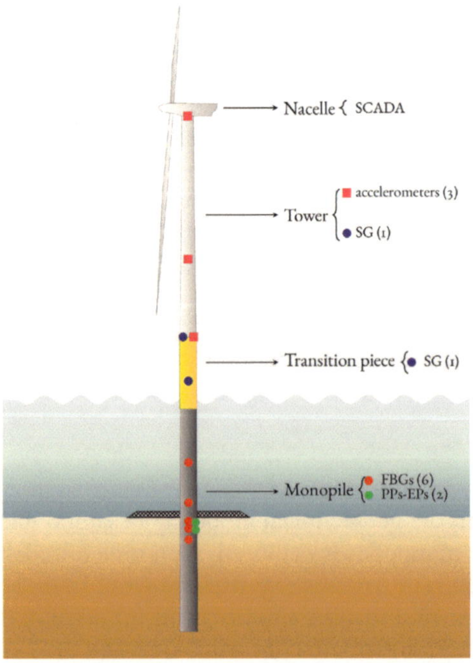

Figure 1.
Instrumented offshore wind turbine structure with accelerometers, strain gauges (SG), fiber Bragg gratings (FBGs), Earth pressure (EP), and pore pressure (PP) sensors.

3. Back-Analysis of Monopile In-Situ Behavior: The final part executes the back-analyses of the monopile's in situ behavior, using the data and methodologies discussed in the previous sections to validate and improve the geotechnical models. The bending moments during quasi-static thrust loading for different wind speeds are back-analyzed for a monitored monopile foundation.

The chapter describes a robust, validated framework for geotechnical modeling that can be applied to improve the design and performance of offshore wind turbine foundations.

2. State-of-art of monopile-soil interaction modeling

The monopile foundation absorbs the combination of bending moment and shear force generated by the combined action of wind, waves, and current on the structure. These actions generate a certain amount of soil reaction. The bending moment profile along the length of the foundation pile depends on the amount of reaction generated by the soil surrounding the pile and the bending stiffness of the pile itself. The behavior of a monopile embedded in a soil mass has been extensively described in the literature through scale model testing and numerical analysis [3–6]. Three modeling approaches were calibrated to such experiments. These modeling approaches are discussed in detail in the subsequent paragraphs.

2.1 3D continuum models

3D continuum models are models in which the soil is modeled as a continuum with the stress-strain relations of each soil element being defined. Soil is a natural material which exhibits complex stress-strain relations which depend on the composition of the soil, the in situ stresses, the magnitude of the applied loads, and the stress history. Soils exhibits volumetric hardening (densification under isotropic pressures) and can exhibit both strain softening and strain hardening during shearing depending on the initial state of the soil. Several researchers have attempted to formulate constitutive models which capture these complex stress-strain relations [7–10]. These constitutive models need to be calibrated to advanced laboratory testing and can then be implemented in a 3D finite element model of the pile and the soil continuum surrounding it. By subjecting the monopile to shear forces and bending moments, the stress redistribution in the soil and the resulting strains can be calculated using a non-linear FE solver. Such calculations are computationally expensive, which prevents them from being used in routine design (although increased adoption for design checks is noticed). Nevertheless, 3D FEA does provide detailed insight into the mobilization of resistance in the soil continuum, and targeted 3D FEA is used to benchmark calculations with more simplified tools [11].

2.2 1D beam-column models with distributed soil springs

1D beam-column models represent the pile through 1D beam elements with non-linear spring representing the soil behavior being defined at each node. While 3D FEA models are not used for routine design, they can be used to calibrate 1D beam column models [11]. These models lump the mobilized soil resistance in a horizontal slice of soil surrounding a given section of the pile into non-linear response curves. The pile is modeled as a one-dimensional beam, discretized into several elements. At each pile node, response curves are applied which capture the development of soil resistance with increasing displacement and cross-sectional rotation of the pile nodes. These response curves depend on the mechanical properties of the soil and the dimensions of the pile. At each pile node, the relation from Eq. (1) must be satisfied where P is the soil's lateral reaction force per unit pile length, y is the pile lateral deflection, z is the vertical coordinate along the pile, and EI is the bending stiffness of the pile.

$$P = EI\frac{d^4y}{dz^2} \tag{1}$$

The response curves are calibrated to instrumented lateral load tests on model piles, most often carried out onshore. Alternatively, the response curves can be derived from scaling laboratory stress-strain curves [12].

For earlier offshore wind farm sites, the only available guidance was derived from Oil & Gas practice as given in the API RP2 GEO standard [13]. Natural frequency measurements on operational offshore wind turbines revealed that the soil stiffness derived with the API RP2 GEO method was too low [4, 14]. Kallehave et al. [15] proposed a modification to the API RP2 GEO response curves to better capture the soil stiffness for operation loads, but this still resulted in underprediction of the pile-soil interaction stiffness, albeit a smaller underprediction compared to API RP2 GEO. For offshore wind turbine monopiles, the response curve formulation from the PISA Joint Industry Project [3] is considered as the most reliable guidance to date, as this formulation offers a better approximation of the soil stiffness across a wide range of load conditions.

The PISA response curves were developed based on onshore field tests in sand (at Dunkirk, FR) and stiff clay (at Cowden, U.K.). Numerical 3D finite element models were calibrated to the field conditions and used to extract the non-linear pile-soil interaction stiffness along the monopiles [3] for a 1D beam-column model. As monopiles have a greater diameter than jacket piles, the P-y curves were supplemented with distributed moment-reaction curves (M-θ curves) which represent the development of a resisting moment with pile cross-section rotation. Monopiles have a lower L/D ratio than jacket piles leading to stiffer behavior and the development of significant shear force (H_B) and moment resistance (M_B) at the pile base. Response curves for these resistance components were also included in the PISA method.

The PISA method was successful in back-analyzing the field experiments at Cowden and Dunkirk by taking the strain-dependent stiffness of the soil into account. The conical response curves which are employed in the PISA method use the small-strain shear stiffness (G_{max}) as an input parameter. This parameter can be derived from in situ testing with the seismic piezocone penetration test (S-PCPT) or from laboratory tests such as bender element or resonant column testing. Location-specific G_{max} measurements are often not available, and the parameter is then derived from conventional CPT data. This introduces a transformation uncertaintly in the parameter estimate.

When a wind turbine structure transfers shear forces and bending moments to the soil, the mobilization of the response curves is calculated using a 1D non-linear finite element algorithm. Eq. (2) is solved in which $\mathbf{F}$ is the vector of applied forces, $\mathbf{u}$ is the displacement vector (containing both translational and rotation terms), and $\mathbf{K}$ is the stiffness matrix. This stiffness matrix is obtained by superposition of the stiffness matrix for the pile, $\mathbf{K}_{\text{pile}}$, and the stiffness matrix of the soil, $\mathbf{K}_{\text{soil}}$. As the soil stiffness matrix is displacement-dependent due to the non-linearity of the soil behavior, the calculation needs to be performed in an iterative manner until the externally applied forces are in equilibrium with the reactions from the mobilized response curves.

$$\mathbf{F} = \mathbf{K} \cdot \mathbf{u} = \left(\mathbf{K}_{\text{soil}} + \mathbf{K}_{\text{pile}}\right) \cdot \mathbf{u} \tag{2}$$

2.3 Lumped stiffness matrices

The most simple model of pile-soil response lumps the pile-soil interaction into a (linearized) stiffness matrix at the mudline Eq. (3) [16]. This stiffness matrix relates the shear force H and bending moment M applied at the mudline to the lateral deflection u and pile head rotation θ at the mudline. The stiffness matrix consists of a lateral stiffness coefficient (K_L), a rotational stiffness coefficient (K_R), and an interaction term (K_{LR}). For monopiles, the rotational stiffness term dominates the response in most cases. The pile response at the mudline and the resulting stiffness coefficients can be calculated by running the more complex beam-column and continuum model types.

$$\begin{bmatrix} H \\ M \end{bmatrix} = \begin{bmatrix} K_L & K_{LR} \\ K_{LR} & K_R \end{bmatrix} \cdot \begin{bmatrix} u \\ \theta \end{bmatrix} \tag{3}$$

This approach allows rapid estimates of the natural frequency of the structure [5], but the formulation with a linear stiffness matrix has a number of limitations. As the increasing soil resistance with depth and the increasing mobilization of deeper layers for high loads is not modeled, the stiffness matrix is only valid for a narrow load range. Soil stratigraphy is not taken into account, which makes the selection of stiffness matrix coefficients in layered soils cumbersome. In general, analytical formulae for the stiffness coefficients as proposed by Arany et al. [5] have limited applicability and do not allow sufficient differentiation between the stiffness properties of the individual foundations in an offshore wind farm for the purpose of foundation response back-analysis.

2.4 Pile-soil interaction modeling: Conclusions

The different modeling approaches are illustrated in **Figure 2**. While 3D finite element models can be computationally expensive and lumped stiffness matrix models are oversimplified, 1D beam-column models provide a trade-off between model complexity and analysis execution time. Calculations are relatively rapid (execution time < 5 seconds on a modern PC), and this type of model is thus used routinely to calculate large amounts of design load cases. The remainder of this chapter will thus focus on the response curve formulations which are employed in these 1D beam-column models.

3. Improvements to pile-soil interaction modeling with 1D beam-column models

3.1 PISA rule-based method

In order to improve the modeling of pile-soil interaction stiffness across a range of load conditions, the PISA rule-based method [17] proposes non-linear response curves

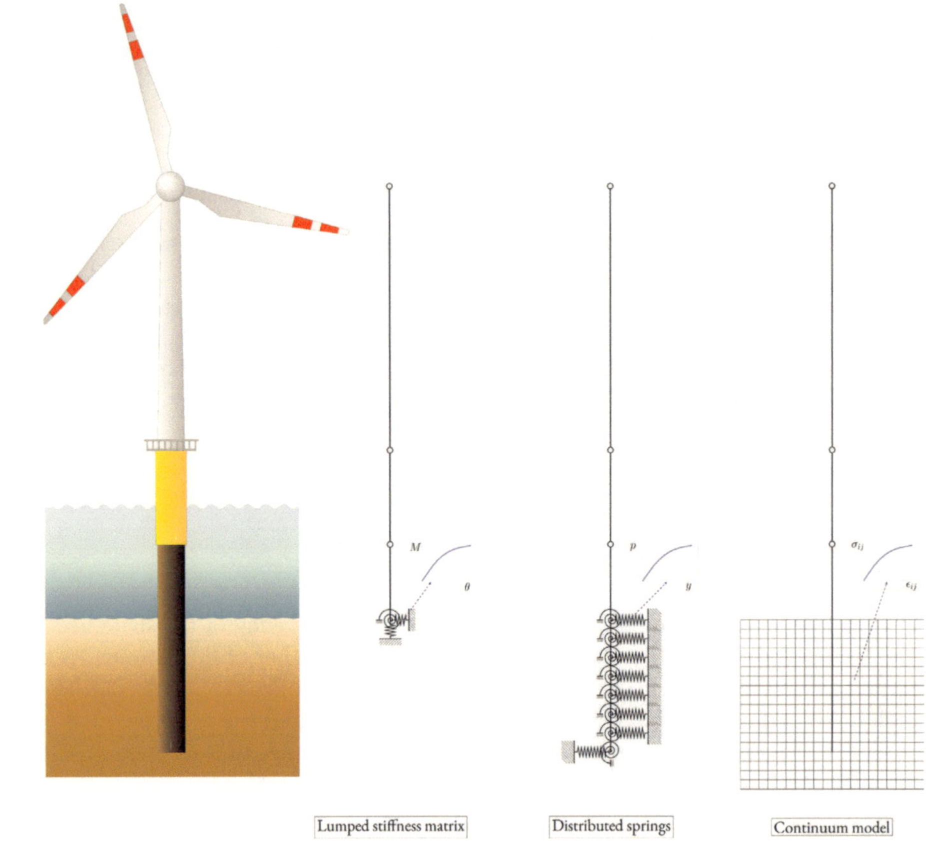

Figure 2.
Modeling approaches for pile-soil structure interaction modeling.

for 1D beam-column modeling of the monopile and surrounding soil. Four types of response curves are proposed by the PISA researchers [18]:

- *P-y* curves describing the mobilization of lateral load with increasing lateral displacement;
- *M-θ* curves describing the mobilization of moment resistance with increasing pile cross-section rotation;
- A H_B-*y* curve describing the mobilization of lateral resistance with increasing pile toe lateral displacement;
- A M_B-*θ* curve describing the mobilization of moment resistance with increasing pile toe cross-section rotation.

A parametric conical response curve is used for each of the response curve types Eq. (4). The response curve has four parameters:

- The ultimate strength of the curve $\overline{\sigma}_u$;
- The strain at the ultimate strength $\overline{\epsilon}_u$;

	$\overline{\sigma}$		$\overline{\varepsilon}$	
	SAND	**CLAY**	**SAND**	**CLAY**
P-y curve	$\frac{P}{S_u D}$	$\frac{P}{\sigma'_{vo} D}$	$\frac{y}{D}\frac{G_{max}}{S_u}$	$\frac{y}{D}\frac{G_{max}}{\sigma'_{vo}}$
M-θ curve	$\frac{M}{S_u D^2}$	$\frac{M}{\lVert P\rVert D}$	$\theta\frac{G_{max}}{S_u}$	$\theta\frac{G_{max}}{\sigma'_{vo}}$
H_B-y curve	$\frac{H_B}{S_u D^2}$	$\frac{H_B}{\sigma'_{vo} D^2}$	$\frac{y}{D}\frac{G_{max}}{S_u}$	$\frac{y}{D}\frac{G_{max}}{\sigma'_{vo}}$
M_B-θ curve	$\frac{M_B}{S_u D^3}$	$\frac{M_B}{\sigma'_{vo} D^3}$	$\theta\frac{G_{max}}{S_u}$	$\theta\frac{G_{max}}{\sigma'_{vo}}$

Table 1.
Response curve normalized force/moment and displacement/rotation terms [18].

- The initial stiffness of the curve k_0;
- The curvature of the curve before reaching the ultimate strength n which controls the degradation of the curve's stiffness for small strains.

$$-n\left(\frac{\overline{\sigma}}{\overline{\sigma}_u}-\frac{\overline{\epsilon}}{\overline{\epsilon}_u}\right)^2+(1-n)\cdot\left(\frac{\overline{\sigma}}{\overline{\sigma}_u}-\frac{\overline{\epsilon}\cdot k_o}{\overline{\sigma}_u}\right)\cdot\left(\frac{\overline{\sigma}}{\overline{\sigma}_u}-1\right)=0 \quad (4)$$

The general response curve from Eq. (4) is denormalised based on the response curve type as shown in **Table 1**. The expressions for normalized displacement or rotation $\overline{\epsilon}_u$, normalized reaction $\overline{\sigma}_u$, initial stiffness k_0 and curvature n are provided by Burd et al. [11] and are not repeated here for the sake of brevity. The denormalisation shown in **Table 1** requires knowledge of the following geotechnical parameters:

- General geotechnical parameters:
  - Effective unit weight γ' for overburden stress calculation.
  - Soil type. Cohesionless soil is denoted as *sand* and cohesive soil as *clay*. Note that the behavior of *silt* depends on the drainage regime. If the silt behaves undrained, it should be modeled as a clay while during drained behavior it will behave as a sand.
- Geotechnical parameters for sand:
  - Small-strain shear modulus G_{max}.
- Geotechnical parameters for clay:
  - Small-strain shear modulus G_{max}.
  - Undrained shear strength S_u. The PISA rule-based method makes a distinction between stiff clay for which the method is calibrated to Cowden clay and soft clay for which the method is calibrated to Bothkennar clay [11]. A comparison of the stress-strain relations of the clay at the monopile location under consideration to the stress-strain response of these reference materials is recommended.

The importance of the small-strain stiffness G_{max} for the development of response curve displacement or rotation is noted. As the monitoring data are typically recorded for episodes in the monopile's lifetime where the loading is lower than the loading in the ultimate limit state (ULS), an accurate charactersation of the development of displacements and rotations for this load regime is required for a meaningful back-analysis. Specific attention is devoted to the selection of the soil stiffness parameter in the subsequent section.

The undrained shear strength S_u characterizes the ultimate strength in clays and can be derived from unconsolidated undrained triaxial (UU), consolidated undrained (CU) triaxial or direct simple shear (DSS) tests on undisturbed samples. Correlations based on the cone penetration test (CPT) can be calibrated on the available tests [19, 20]. Because the ultimate strength is rarely reached, the selection of S_u is not discussed further.

3.2 Determination of G_{max} at monopile foundation locations

3.2.1 In-situ measurements with the seismic piezocone penetration test

As the CPT is a destructive test involving significant deformation of the soil, the stiffness at limited deformations cannot directly be derived from the CPT data. To allow direct measurement of the soil behavior at small strain using the same equipment as a conventional CPT test, the S-PCPT test was developed. Two sets of additional accelerometer sensors (geophones) are placed on the cone rod to allow the detection of seismic waves. Seismic shear waves (S-waves) are then generated at the seabed using a seismic source placed at a known offset to the CPT rod. This setup is known as the S-PCPT. The acceleration of the geophone sets is measured, and by calculating the cross-correlation between the signals received at the upper and lower geophone, the difference in arrival time can be identified. When the offset between the two geophone sets is divided by the difference in arrival time, the seismic shear wave velocity V_s is obtained.

The seismic shear wave velocity (V_s) has a direct relation to the small-strain shear modulus (G_{max}) when the bulk density of the soil (ρ) is known Eq. (5). It should be noted that the small-strain shear modulus can exhibit anisotropy [21] with differences between the value in horizontal and vertical direction. The S-PCPT test measures accelerations in both vertical and horizontal directions with the horizontal component being more relevant for pile-soil interaction. Furthermore, S-PCPT measurements near the seabed can be affected by noise due to the interference between surface waves and shear waves. To increase the accuracy of the S-PCPT, a system with dual geophones, as described above, is preferred. Systems with a single geophone suffer more from noise as the arrival time of the shear wave needs to be selected from the noisy data. Despite the measurement uncertainty described above, the S-PCPT test is the most commonly used test to determine the small-strain shear modulus of uncemented sedimentary soils in situ.

$$G_{max} = \rho V_s^2 \tag{5}$$

3.2.2 Indirect determination from G_{max} from CPT-based correlations

The S-PCPT tests described in the preceding sections are usually not available at every wind turbine foundation. However, a site-specific profile of small-strain shear modulus is required to perform pile response calculations with the PISA method. Correlations between conventional CPT measurements and small-strain shear moduli are used in such cases. Each of these correlations is based on a unique dataset combining direct measurements of shear wave velocity (or small-strain shear modulus) with CPT testing at the same location and depth. The datasets contain measurements in specific soil types and the proposed correlations apply to the in situ stress conditions at the sites used for calibration. Using such correlations introduces a model uncertainty or transformation uncertainty [22] in the geotechnical analysis as the small-strain shear modulus is derived indirectly.

As the correlation used in geotechnical design has a direct impact on the results of the geotechnical analysis, the correlations which are most often used in practice are critically reviewed by Stuyts et al. [23] for application to North Sea soil conditions. In that study, the estimates of G_{max} or V_s with CPT-based correlations which are routinely used in monopile geotechnical design are compared against a database of S-PCPT measurements from the Belgian, Dutch, German, and Danish offshore sector. The ratio of predicted G_{max} or V_s to the measured values was calculated for each S-PCPT measurement, and the performance of each correlation was summarized through the mean and coefficient of variation (COV) of these ratios across the dataset. A correlation performs well if it has a mean close to one (neutral bias) and a low COV. In the study by Stuyts et al. [23], a new correlation with a power law dependence on the normalized vertical effective stress σ_{v0}' Eq. (6) showed improved performance against the available North Sea dataset, with a nearly neutral bias and the smallest coefficient of variation as shown in **Table 2**. With this correlation, location-specific G_{max} profiles can be derived at each CPT location, and the effect of changes in the overburden on G_{max} can be calculated.

$$V_s = 10^{2.075-0.213 \cdot I_c} \left(\frac{\sigma'_{vo}}{1\text{kPa}} \right)^{0.77+0.25 \cdot \log_{10}(2.075-0.213 \cdot I_c)} \tag{6}$$

Correlation	Predicted property	N	μ	COV	R^2
Hardin and Black [24]	G_{max}	3058	1.570	0.652	−2.414
Rix and Stokoe [25]	G_{max}	2656	0.809	0.713	−0.006
Mayne and Rix [26]	G_{max}	237	1.213	0.742	−4.441
Peuchen et al. [27]	G_{max}	3058	0.846	0.733	−0.017
Wride et al. [28]	V_s	3058	0.782	0.214	−0.984
Hegazy and Mayne [29]	V_s	3058	1.131	0.260	−0.949
Andrus et al. [30]	V_s	3058	1.095	0.207	−0.065
Tonni and Simonini [31]	V_s	863	2.320	0.550	−76.080
Robertson and Cabal [20]	V_s	3058	1.056	0.206	0.155
McGann et al. [32]	V_s	3058	0.901	0.207	0.036
Stuyts et al. [23]	V_s	3058	1.007	0.188	0.370

Table 2.
Accuracy of selected CPT-based correlations for V_s and G_{max} on the North Sea dataset in terms of the mean (μ) and coefficient of variation (CoV) of the ratio of calculated to measured G_{max} or V_s. The R^2 score is also shown for each correlation based on the N available S-PCPT measurements.

3.3 Effect of scour protection

When offshore wind turbine monopiles are embedded in soil profiles with cohesionless soils at the seabed, scour may occur, resulting in a local lowering of the seabed. Such seabed lowering would increase the bending moments on the monopiles, and thus a scour protection layer consisting of stones is placed before or after foundation installation. The scour protection has an extent of 1.5 to 2 times the monopile diameter from the monopile wall. Depending on the global sediment transport regime, either an exposed situation (with the scour protection protruding above the seabed) or a submerged situation (with the scour protection below the global seabed level) can develop over time. After installation on the top of the virgin seabed, the scour protection may sink into the seabed by a certain amount. Sufficient rock material needs to be dumped to ensure that the rock reaches a certain level defined by the wind farm operator after the installation. This scour protection layer increases the vertical effective stresses in the shallow layers which can in turn increase the small-strain shear modulus as shown in Eq. 6. The scour protection may also provide additional lateral fixity to the pile. As there is no generally accepted methodology to quantify the presence of the scour protection system in a monopile lateral response analysis, three different analysis methods are shown in **Figure 3**:

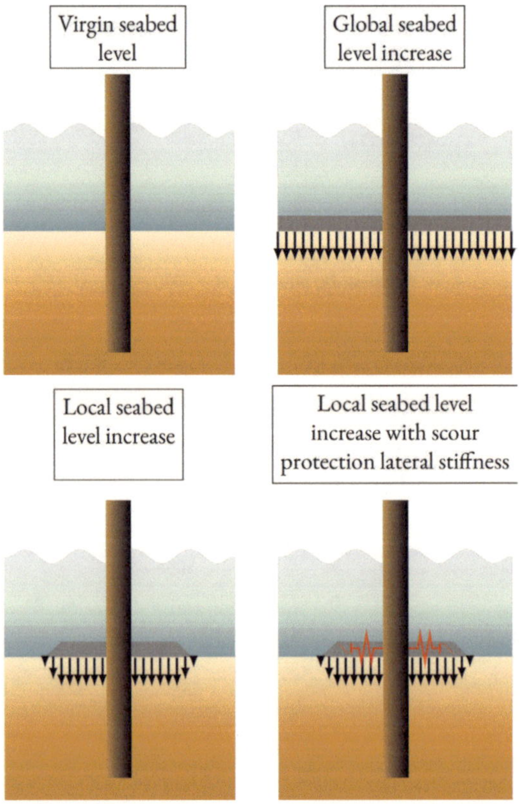

Figure 3.
Modeling approaches for quantifying the effect of the scour protection on the pile-soil interaction.

- Virgin seabed level: In this approach, the presence of the scour protection layer is completely disregarded.

- Global seabed level increase: The effect of the scour protection layer is quantified by adding a layer with the thickness of the scour protection system on the top of the surface layer.

- Local seabed level increase: A more accurate approach is to calculate the stress changes due to scour protection with a finite extent using elastic stress formulae. The elastic stress formulae are obtained by integrating the formula proposed by Boussinesq [33] for the stress increase due to a point load, over the area of the seabed affected by the scour protection.

- Local seabed level increase with scour protection lateral stiffness: An additional modeling approach can be defined in which the increase in vertical effective stress obtained from elastic stress formulae and G_{max} increases according to Eq. (6) are applied and lateral pile-soil interaction stiffness is also modeled in the scour protection layer.

The effect of the four modeling approaches on the vertical effective stress and the small-strain shear modulus profile for a selected wind turbine location is shown in **Figure 4**. For the case with virgin seabed, the vertical effective stress and small-strain shear modulus are not defined above the sand layer. The case with a global seabed level increase leads to an increase in vertical effective stress which applies over the entire depth range. In this case, a small but non-zero stiffness is assigned to the scour protection layer. When the localized nature of the scour protection layer is considered, the vertical effective stresses are still increased by the presence of the scour protection, but the increase reduces with depth until, at around 25 m depth, the presence of the scour protection is no longer felt by the soil. Because of the increase in overburden stress, an increase in G_{max} is

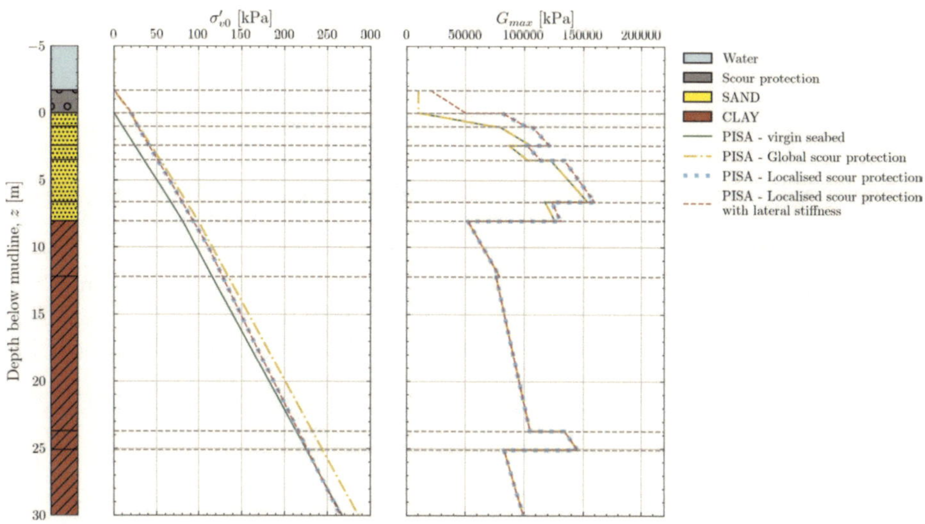

Figure 4.
Vertical effective stress and small-strain shear modulus profiles for different analysis cases at a selected wind turbine location.

also considered according to Eq. (6). It should be noted that there is no generally accepted guidance for modeling the lateral support from the scour protection. For the final case, the lateral stiffness of the scour protection was modeled with the PISA formulation by assuming a stiffness equivalent to the stiffness of a dense sand at shallow depths.

The effect of the different approaches for modeling scour protection is checked against on the in situ monitoring data in the next section.

4. Back-analyses of monopile in situ behavior

4.1 Bending moments from strain measurements

The bending moments in a steel tubular pile can be obtained by measuring the strain in the pile wall. Eq. (7) provides the fundamental relation between bending strain (ϵ) and the bending moment (M). This requires the distance y between the location of the strain measurement and the neutral axis of the pile to be known. The second moment of area (I) also needs to be known. When the outer diameter (D_o) and the wall thickness of the pile are known, the calculation of the internal diameter (D_i) is straightforward.

$$\epsilon = -\frac{My}{EI} \qquad \text{where } I = \frac{\pi}{64}\left(D_o^4 - D_i^4\right) \tag{7}$$

Eq. (7) shows the relation between bending moment and strain when the strain is measured or calculated in the plane of the applied bending moment. For offshore wind turbine structures, the direction of the bending varies with the wind direction, so a minimum of three measurement points are required to calculate the bending moment aligned with the direction of the wind, M_{tn}, the bending moment perpendicular to the wind direction, M_{tl} and the normal force, N.

Two different measurement techniques are used for strain measurements on offshore wind turbine monopiles. As shown in **Figure 1**, strain gauges are installed above the water level. A temperature sensor (thermocouple) is installed next to each strain gauge to allow for thermal compensation of the measurements. Strain measurements below the water level and below the mudline are made possible by optical fibers with Fiber Bragg Gratings (FBG); refractive pattern which are inscribed on the fiber [34]. Strains cause a change of the reflected wavelength of the FBG which can be measured.

The instrumented monopiles are equipped with four optical fibers at 90° intervals along the pile circumference as shown in **Figure 5**. The optical fibers are subjected to very high dynamic stresses during pile driving which requires careful attachment of the fiber to the monopile. Moreover, the embedded section of the optical fibers is subjected to shear stresses as it passes through the soil. This prevents the use of FBGs near the pile tip, as field experience shows that these sensors do not survive the installation. Most of the FBGs used in this work are located relatively close to the mudline, where the highest bending moments are expected.

4.2 Data processing

The strain data are first processed to remove zero offsets. It should be noted that the driving process can introduce a non-zero initial strain in the FBG sensors [35].

Figure 5.
Location of optical fibers with FBGs on the inside of an instrumented monopile [2].

To identify the zero offsets of the sensors, the strain signals are checked as the wind rotor yaw angle varies. If the rotor axis is aligned with the sensor direction, the maximum strain is expected. If the rotor axis is oriented perpendicular to the sensor direction, a zero crossing should be noted. Henkel [35] noticed that the sensors can drift over time. A recommendation was formulated to recalculate the offsets once a month.

When strain measurements are available at 90° intervals along the monopile circumference, zero-offsets and temperature compensation is first applied to the signals. The strains are then used to calculate bending moments along the N-S and E-W axes (M_x and M_y) according to Eq. (7) using a least-squares solver. The normal force N is also solved. These bending moments can be converted into the wind turbine nacelle's frame of reference which requires knowledge of the wind turbine yaw angle from SCADA data. The direction aligned with the wind turbine axis is called the fore-aft (FA) direction, and the strains in this direction are influenced by the quasi-static thrust load on the rotor. The direction perpendicular to the wind turbine axis is called the side-side (SS) direction, and the strains in this direction are mainly influenced by the first-order mode.

The strains and bending moments are available for a 10-min time series with a 100 Hz sampling frequency. An example of a 10 min time series of calculated bending moments in the FA and SS direction is shown for a selected episode during the wind turbine operation in **Figure 6**. The signals show a non-zero quasi-static bending moment in the FA direction with oscillations around this value. In the SS direction, bending moments oscillate around a near-zero average as there is no steady loading in the SS direction. The quasi-static bending moment increases along the unsupported section of the monopile (above -30mLAT). Below the mudline, the quasi-static bending moment reduces as the moment is transferred to the subsoil. A closer inspection of the three uppermost traces at 17mLAT, 5mLAT, and -17mLAT show a time lag between the signals. These signals are connected to different data acquisition systems for which the clocks are not synchronized. Henkel [35] proposes a correction whereby the cross-correlation between the signals is taken to identify the offset. This correction was not applied to the data shown in **Figure 6**.

The bending moment in the fore-aft and side-side direction can be calculated for the monitored lifetime of an offshore wind turbine monopile as shown in **Figure 7**. The 10-minute averages are normalized by the maximum bending moment in the fore-aft direction and color-coded with the rotor speed. The figure shows that the

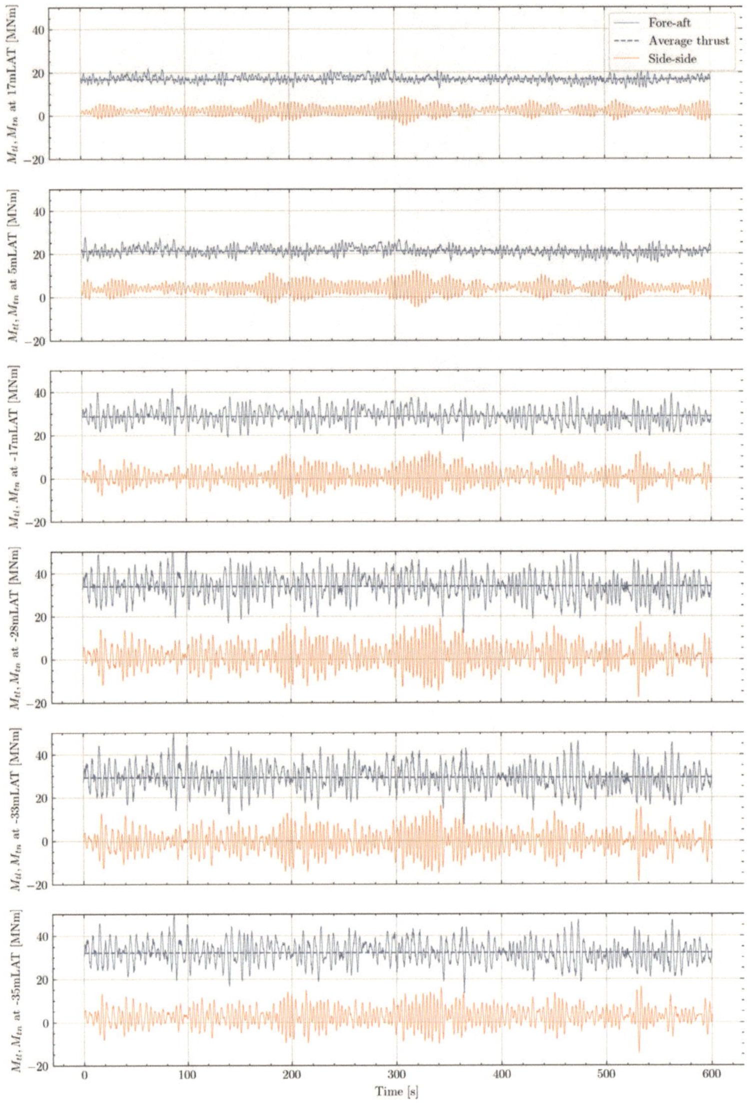

Figure 6.
Example of a 10-min time series of bending moments in the FA and SS direction for six different sensor elevations. The top two elevations correspond to strain gauge measurements and the bottom four to FBG measurements. The quasi-static bending moment in the FA direction is shown as a dashed line.

bending moments in the fore-aft direction increase with increasing wind speed when the turbine is rotating. This is because the thrust load of the wind on the rotor blades increases. The bending moments increase until the rated power of the wind turbine is reached. At this point, the bending moments in the fore-aft direction reach a maximum. When the wind speed increases further, the wind turbine blades are pitched, leading to a reduction in thrust load. In the side-side direction, the bending moments are much smaller due to the lack of thrust load in that direction. When the rotor speed is low, the bending moments in the fore-aft direction and side-side direction are of similar magnitude. For such conditions, the bending moments are only generated by the environmental loading.

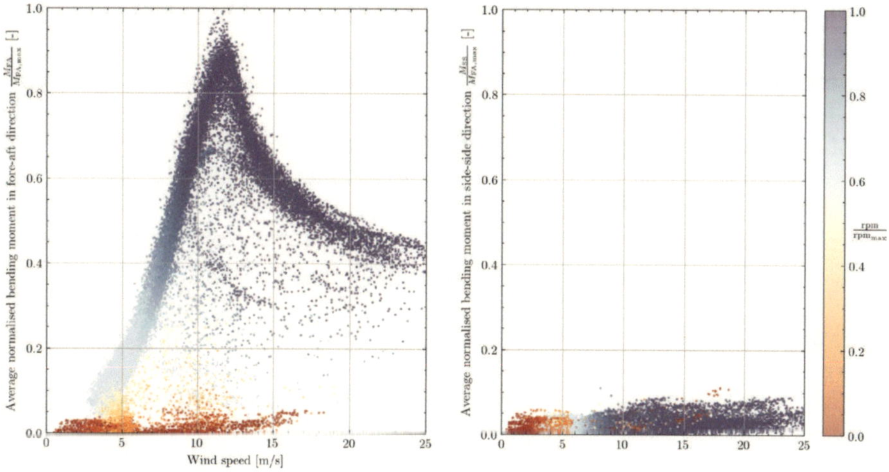

Figure 7.
Normalized 10-minute averages of bending moments in the fore-aft (left panel) and side-side direction (right panel) for various wind and rotor speeds.

The quasi-static bending moment in the FA direction is especially useful for verifying lateral pile response calculation methodologies. To obtain quasi-static thrust loads, episodes with steady operational and metocean conditions are selected, and the average bending moment in the FA direction is computed. **Figure 6** is an example of a period with steady conditions. The signals show stationarity and a reliable estimate of the quasi-static bending moment is obtained. This process can be repeated for several operational conditions.

Figure 8 shows the normalized measured bending moments at the strain gauge and FBG sensor elevations. The bending moment increases until the soil starts providing lateral support to the monopile, after which the bending moments are further absorbed with increasing depth. The maximum bending moment is observed above

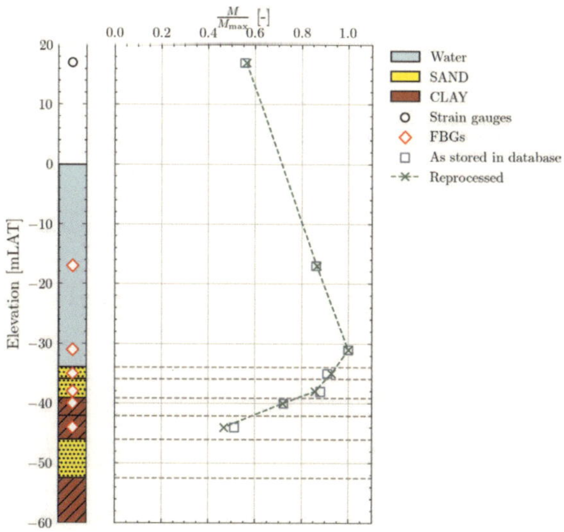

Figure 8.
Normalized measured quasi-static bending moment profile for at a selected monitored location in a Belgian offshore wind farm. The correction from updated sensor offset is shown with the green-dashed line.

the virgin mudline. This can be attributed to the presence of scour protection as discussed in the next sections.

The monitoring data available for this study were processed with fixed sensor offsets. However, sensor drift requires periodic recalibration of the sensors [35]. The effect of this is visible in **Figure 8**. The original data show significant deviation from the expected trend for the first point below the mudline (−35mLAT) where a sudden drop of the bending moment is observed which cannot be related to the presence of a competent soil layer at this elevation. After correction for sensor drift, the monitored bending moments show a smoother profile which is in line with expectations.

4.3 Bending moment back-analysis

The 1D beam-column model described in Section 2 was applied for a permanently monitored location to check the impact of several modeling choices on the correspondence between measured and calculated bending moments. First, the effect of the response curve method is checked. The API RP2 GEO method (with and without modification according to Kallehave et al. [15]) and PISA rule-based methods are compared. Next, the effect of seabed level changes due to the presence of a scour protection is examined. Several methods for quantifying the effect of the scour protection are discussed.

4.3.1 Monitored location

The bending moment measurements discussed in this chapter are calculated from strain data at an offshore wind turbine structure equipped with a permanent monitoring system. One strain gauge level and six FBG levels were functional for the selected monopile. The monopile tip depth is located at 29.0 m below the virgin seabed level. The selected WTG location has a dense sand layer of 6 m thickness at the surface overlying a layer of stiff overconsolidated clay. Below the clay layer, another dense sand layer of 6 m thickness is found. The monopiles tip depth is located in a stiff clay layer. A scour protection layer with a total thickness of 1.7 m is installed around the monopile.

Three different episodes from the available monitoring data were selected for back-analysis:

- Wind speed = 6.5 m/s: wind speed below the wind speed at maximum rated power
- Wind speed = 11.6 m/s: wind speed at maximum rated power
- Wind speed =14.7 m/s: wind speed above the wind speed at maximum rated power

4.3.2 Determination of shear force and bending moment

Estimating the shear force and bending moment at the mudline is usually performed with advanced load estimation software which quantifies the effect of wind, wave, current, and inertial loads and performs coupled aero-hydro-servo elastic simulation in the time domain [36]. Such load estimation is beyond the scope of this work, and therefore, a pragmatic approach was followed to derive the combination of quasi-static shear force and bending moment to be used in back-analyses. The bending moments inferred from strain measurements were used to determine a horizontal force which leads to the expected increase in the bending moment between two selected elevations z_1 and z_2 as shown in Eq. (8). As bending moments measurements are available above the mudline from strain gauge and FBG data, these measurements can be used to determine the horizontal force which satisfies this equation. This equation assumes that no quasi-static external loading is imposed on the pile in the region between the selected sensor elevations and the mudline. This is an approximation as the quasi-static current loading in this region will not be equal to zero. Nevertheless, the current loading will be much lower than the applied quasi-static thrust load, and representative loads were obtained with this procedure.

$$M(z_2) = M(z_1) + H \cdot \|z_2 - z_1\| \tag{8}$$

4.3.3 Effect of response curve modeling

The influence of the response curve method on the bending moment profile is investigated first for the three selected load levels. The results for the selected location is provided in **Figure 9**. The bending moment profiles obtained for the API RP2 GEO method, Kallehave et al. method, and PISA method are plotted together with the bending moments inferred from the strain measurements for the three selected load levels.

It should be noted that the monitored bending moment at -35mLAT deviates significantly from the surrounding points. The recalculation of the offsets was not possible for all monitored episodes. As shown before in **Figure 8**, such a recalibration would bring the points at -35mLAT more in line with the calculated bending moment profiles.

When comparing the different response curve methods, the most pronounced effect is the overprediction of the monitored bending moment with the API RP2 GEO method. This method predicts a decay of bending moment with depth which is clearly too slow. The Kallehave et al. and PISA methods, which are modified to capture the higher stiffness of the soil at small strains, provide a closer match to the monitoring data. The trend of bending moment dissipation with depth is followed closely by these methods, although an overprediction is still noted, especially for the highest load level.

4.3.4 Influence of scour protection modeling

The influence of the scour protection is checked subsequently. The results for the selected monitored location is provided in **Figure 10**. The influence of the scour

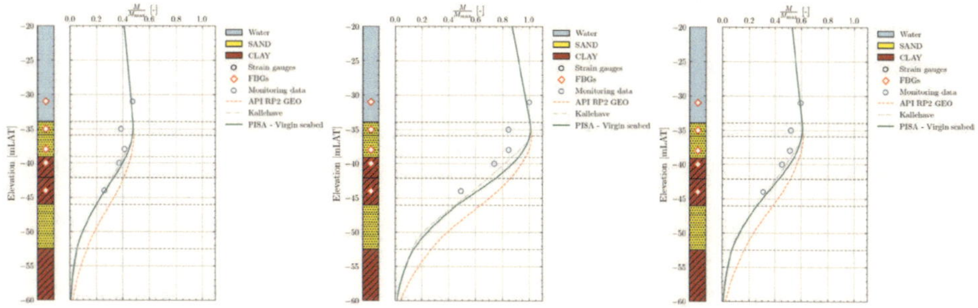

(a) Wind speed = 6.5m/s (b) Wind speed = 11.6m/s (c) Wind speed = 14.7m/s

Figure 9.
Influence of the response curve method on the back-analysis results.

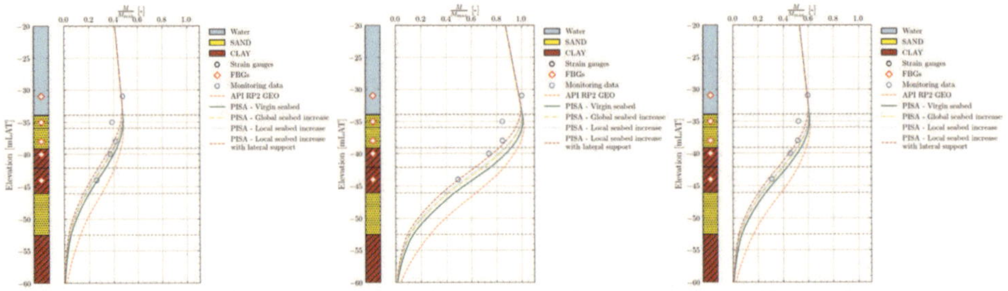

(a) Wind speed = 6.5m/s (b) Wind speed = 11.6m/s (c) Wind speed = 14.7m/s

Figure 10.
Influence of the scour protection modeling strategy on the back-analysis results.

protection was only checked for the PISA method as this method allowed taking the stress dependence of G_{max} into account.

Adding the effect of the scour protection leads to even more rapid bending moment decay with depth and generally brings the back-analysis results closer to the measurements. At the selected monitored location, the scour protection is 1.7 m thick, which leads to a pronounced effect. When inspecting the results, the effect of adding a global scour protection layer with almost no lateral stiffness appears to lead to similar results as a more detailed modeling of the localized effects of the scour protection.

Overall, the method which leads to the closest overall match between measurements and back-analysis is the one where the effect of scour protection on the vertical effective stress and G_{max} is modeled together with a lateral stiffness of the scour protection itself. Even though the lateral stiffness of the scour protection layer was selected on a fairly arbitrary basis, the results indicate that it is likely that there is a lateral stiffness contribution from the scour protection.

5. Conclusions

In situ monitoring data from instrumented offshore wind turbine locations can be used to validate geotechnical design methods which were developed based on scale model tests and numerical analysis. An accurate prediction of the pile-soil interaction stiffness is important for fatigue calculations for offshore monopiles. Monitoring of bending moments during quasi-static thrust loading can establish confidence in the selected pile-soil interaction stiffness.

The available bending moment monitoring data offer essential insights into the dissipation of bending moments in the monopile with depth. Strain measurements below the mudline show how the different soil layers provide reaction against the loading imposed by the quasi-static thrust load. The lateral and moment-rotational reaction provided by the soil can be verified for full-scale monopiles during their operational lifetime.

Based on the bending moment back-analysis, it can be concluded that the PISA rule-based method provides an applicable framework for offshore wind turbine monopiles subjected to the available range of thrust loads, if the effects of scour protection on the stress and stiffness of in the subsoil are accounted for. Modeling some degree of lateral stiffness for the scour protection appears to be necessary, but the theoretical or experimental basis for doing so is currently absent.

The pile-soil interaction stiffnesses which are validated based on the bending moment back-analysis can be carried over to an integrate structural model of the entire offshore wind turbine structure [37]. The first natural frequency can also be well approximated with these stiffnesses, providing further confidence in the proposed geotechnical modeling strategy.

Acknowledgements

The authors would like to acknowledge the support of the Belgian Ministry of Economic Affairs through the ETF project WINDSOIL project. The support of VLAIO through the De Blauwe Cluster SBO SOILTWIN project is also acknowledged. Geotechnical data from the Dutch and German offshore wind farms were used under a Creative Commons License.

Abbreviations

CPT	cone penetration test
EP	earth pressure sensor
FA	fore-aft direction
FBG	fiber Bragg Grating
PP	pore pressure sensor
SG	strain gauge
S-PCPT	seismic piezocone penetration test
SS	side-side direction
ULS	ultimate limit state

Author details

Bruno Stuyts[1,2*†], Wout Weijtjens[1], Carlos Sastra Jurado[1,2†] and Christof Devriendt[1]

1 Vrije Universiteit Brussel, OWI-Lab, Pleinlaan, Brussels, Belgium

2 UGent, Labo Geotechniek, Technologiepark, Zwijnaarde, Belgium

*Address all correspondence to: bruno.stuyts@vub.be

† These authors contributed equally.

References

[1] WindEurope. Offshore wind in Europe: Key trends and statistics 2019. Technical report. 2020

[2] Weijtjens W, Voet E, de Sitter G, Luyckx G, Devriendt C. Fiber Optic Sensors on Monopiles for Sub-Soil Load Monitoring of Offshore Wind Turbines. TU Braunschweig; 2018. pp. 393-403

[3] Byrne BW, McAdam RA, Burd HJ, Houlsby GT, Martin CM, Beuckelaers WJAP, et al. Pisa: New design methods for offshore wind turbine monopiles. In: Offshore Site Investigation Geotechnics Eighth International Conference Proceeding. Vol. 142. London, United Kingdom: Society for Underwater Technology; 2017. pp. 142-161

[4] Kallehave D, LeBlanc Thilsted C, Liingaard MA. Modification of the API PY formulation of initial stiffness of sand. In: Offshore Site Investigation and Geotechnics: Integrated Technologies-Present and Future. London, United Kingdom: OnePetro; 2012

[5] Arany L, Bhattacharya S, Macdonald JHG, John Hogan S. Closed form solution of eigen frequency of monopile supported offshore wind turbines in deeper waters incorporating stiffness of substructure and SSI. Soil Dynamics and Earthquake Engineering. 2016;**83**: 18-32

[6] Page AM, Næss V, De Vaal JB, Eiksund GR, Nygaard TA. Impact of foundation modelling in offshore wind turbines: Comparison between simulations and field data. Marine Structures. 2019;**64**:379-400

[7] Taborda DMG, Zdravković L, Kontoe S, Potts DM. Computational study on the modification of a bounding surface plasticity model for sands. Computers and Geotechnics. 2014;**59**:145-160

[8] Tom Schanz PA, Vermeer P, Bonnier P. The hardening soil model: Formulation and verification. In: Beyond 2000 in Computational Geotechnics. Abingdon, United Kingdom: Routledge; 2019. pp. 281-296

[9] Mašín D. Modelling of soil behaviour with hypoplasticity. In: Springer Series in Geomechanics and Geoengineering. Switzerland AG: Springer Nature; 2019. DOI: 10.1007/978-3-030-03976-9

[10] Liu H, Diambra A, Abell JA, Pisanò F. Memory-enhanced plasticity modeling of sand behavior under undrained cyclic loading. Journal of Geotechnical and Geoenvironmental Engineering. 2020;**146**(11):04020122

[11] Burd HJ, Abadie CN, Byrne BW, Houlsby GT, Martin CM, McAdam RA, et al. Application of the Pisa design model to monopiles embedded in layered soils. Géotechnique. 2020;**70**(11):1067-1082

[12] Zhang Y, Andersen KH. Scaling of lateral pile PY response in clay from laboratory stress-strain curves. Marine Structures. 2017;**53**:124-135

[13] American Petroleum Institute. API RP 2GEO geotechnical and foundation design considerations. Technical report. 2011. Standard

[14] Devriendt C, Weijtjens W, El-Kafafy M, De Sitter G. Monitoring resonant frequencies and damping values of an offshore wind turbine in parked conditions. IET Renewable Power Generation. 2014. ISSN 1752-1416, 1752-1424;**8**(4):433-441. DOI: 10.1049/iet-rpg.2013.0229

[15] Kallehave D, LeBlanc C, Liingaard MA. Modification of the API P-Y formulation of initial stiffness of sand. In: Integrated Geotechnologies - Present and Future. London: Society for Underwater Technology; 2012. pp. 465-472

[16] Gazetas G, Tassoulas JL. Horizontal stiffness of arbitrarily shaped embedded foundations. Journal of Geotechnical Engineering. 1987;**113**(5):440-457

[17] Byrne BW, Houlsby GT, Burd HJ, Gavin KG, Igoe DJP, Jardine RJ, et al. Pisa design model for monopiles for offshore wind turbines: Application to a stiff glacial clay till. Géotechnique. 2020; **70**(11):1030-1047

[18] Beuckelaers W. Numerical modelling of laterally loaded piles for offshore wind turbines [PhD thesis]. University of Oxford. 2017

[19] Lunne T, Powell JJM, Robertson PK. Cone Penetration Testing in Geotechnical Practice. Boca Raton, Florida, United States: CRC Press; 2002

[20] Robertson P, Cabal KL. Guide to cone penetration testing. Technical report. 2015

[21] Wang YH, Mok CM. Mechanisms of small-strain shear-modulus anisotropy in soils. Journal of Geotechnical and Geoenvironmental Engineering. 2008; **134**(10):1516-1530

[22] Phoon K-K, Tang C. Characterisation of geotechnical model uncertainty. Georisk: Assessment and Management of Risk for Engineered Systems and Geohazards. 2019;**13**(2):101-130

[23] Stuyts B, Weijtjens W, Jurado CS, Devriendt C, Kheffache A. A critical review of cone penetration test-based correlations for estimating small-strain shear modulus in north sea soils. Geotechnics. 2024. ISSN 2673-7094;**4**(2): 604-635. DOI: 10.3390/geotechnics 4020033. Available from: https://www.mdpi.com/2673-7094/4/2/33

[24] Hardin BO, Black WL. Vibration modulus of normally consolidated clay. Journal of the Soil Mechanics and Foundations Division. 1968;**94**(2):353-369. DOI: 10.1061/JSFEAQ.0001100

[25] Rix GJ, Stokoe KH. Correlation of initial tangent modulus and cone penetration resistance. In: Calibration Chamber Testing. New York: Elsevier; 1991. pp. 351-362

[26] Mayne PW, Rix GJ. Gmax-qc relationships for clays. Geotechnical Testing Journal GTJODJ. 1993;**16**(1): 54-60

[27] Peuchen J, Kaltekis K, Klein M, Murali M, Erp F, Hicks M. Characteristic Values for Geotechnical Design of Offshore Monopiles in Sandy Soils - Case Study. 10. 2020

[28] Wride CE, Robertson PK, Biggar KW, Campanella RG, Hofmann BA, Hughes JMO, et al. Interpretation of in situ test results from the canlex sites. Canadian Geotechnical Journal. 2000;**37** (3):505-529

[29] Hegazy YA, Mayne PW. A global statistical correlation between shear wave velocity and cone penetration data. In: Site and Geomaterial Characterization. 2006. pp. 243-248

[30] Andrus RD, Mohanan NP, Piratheepan P, Ellis BS, Holzer TL. Predicting shear-wave velocity from cone penetration resistance. In: Proceedings of the 4th International Conference on Earthquake Geotechnical Engineering. Vol. 2528. Thessaloniki, Greece; 2007

[31] Tonni L, Simonini P. Shear wave velocity as function of cone penetration test measurements in sand and silt mixtures. Engineering Geology. 2013; **163**:55-67

[32] McGann CR, Bradley BA, Jeong S. Empirical correlation for estimating shear-wave velocity from cone penetration test data for banks peninsula loess soils in canterbury, new zealand. Journal of Geotechnical and Geoenvironmental Engineering. 2018; **144**(9):04018054

[33] Budhu M. Soil Mechanics and Foundations. Vol. 1. Hoboken, New Jersey, United States: Wiley; 2020

[34] FBGS. FBG principle. 2022. Available from: https://fbgs.com/technology/fbg-principle/ [Accessed: December 28, 2022]

[35] Henkel M, Weijtjens W, Devriendt C. Fatigue stress estimation for submerged and sub-soil welds of offshore wind turbines on monopiles using modal expansion. Energies. 2021; **14**(22):7576

[36] NREL. Fast. 2022. Available from: https://www.nrel.gov/wind/nwtc/fast.html [Accessed: January 17, 2023]

[37] Sastre Jurado C, Stuyts B, Weijtjens W, Devriendt C. Impact of calibrated soil-monopile-interaction model on resonance frequencies. Journal of Physics: Conference Series. 2022;**2265**(3):032098. DOI: 10.1088/1742-6596/2265/3/032098